Samuel Laing

Modern Science and Modern Thought

Samuel Laing

Modern Science and Modern Thought

ISBN/EAN: 9783337034177

Printed in Europe, USA, Canada, Australia, Japan

Cover: Foto ©Thomas Meinert / pixelio.de

More available books at **www.hansebooks.com**

MODERN SCIENCE

AND MODERN THOUGHT

BY

S. LAING.

SIXTH THOUSAND.

CONTAINING A SUPPLEMENTAL CHAPTER ON

GLADSTONE'S "DAWN OF CREATION" AND "PROEM OF GENESIS,"

AND ON

DRUMMOND'S "NATURAL LAW IN THE SPIRITUAL WORLD."

LONDON: CHAPMAN AND HALL, Limited.
1889.

Printed from Stereotype,
March, 1889.

INTRODUCTION TO SIXTH EDITION.

THE Fifth Edition of this work having been exhausted in three weeks from its publication, a Sixth Edition is now published at the same low price of 3s. 6d. The Author's object is to meet the rapidly-increasing demand for works giving an intelligible and popular view of the leading questions of the day in Science and Philosophy, by bringing out the volume at a price which renders it accessible to the general public, and specially to the class of intelligent working men who take an interest in such subjects.

S. LAING.

HALL GROVE, BAGSHOT,
    Aug. 15, 1888.

PREFACE TO FIRST EDITION.

THE object of this book is to give a clear and concise view of the principal results of Modern Science, and of the revolution which they have effected in Modern Thought. I do not pretend to discover fresh facts or to propound new theories, but simply to discharge the humbler though still useful task of presenting what has become the common property of thinking minds, in a popular shape, which may interest those who lack time and opportunity for studying special subjects in more complete and technical treatises.

I have endeavoured also to give unity to the subjects treated of, by connecting them with leading ideas: in the case of Science, that of the gradual progress from human standards to those of almost infinite space and duration, and the prevalence of law throughout the universe to the exclusion of supernatural interference; in the case of Thought, the bearings of these discoveries on old creeds and philosophies, and on the practical conduct of life. The endeavour to show how much of religion can be saved from the shipwreck of theology has been the main object of the second part. Those who are acquainted with the scientific literature of the day will at once see how much J have been indebted to Darwin, Lyell, Lubbock, Huxley, Proctor, and other well-known writers. In fact, the first part of this book does not pretend to be more than a compendious popular abridgment of their works. I prefer, therefore, acknowledging my obligations to them once for all, rather than encumbering each page by detailed references.

The second part contains more of my own reflections on the important subjects discussed, and must stand or fall on its own merits rather than on authority. I can only say that I have endeavoured to treat these subjects in a reverential spirit, and that the conclusions arrived at are the result of a conscientious and dispassionate endeavour to arrive at "the truth, the whole truth, and nothing but the truth."

CONTENTS.

Part I.

MODERN SCIENCE.

CHAPTER I.

SPACE 3

Primitive Ideas—Natural Standards—Dimensions of the Earth—Of Sun and Solar System—Distance of Fixed Stars—Their Order and Size—Nebulæ and other Universes—The Telescope and the Infinitely Great—The Microscope and the Infinitely Small—Uniformity of Law—Law of Gravity—Acts through all Space—Double Stars, Comets, and Meteors—Has acted through all Time.

CHAPTER II.

TIME 20

Evidence of Geology—Stratification—Denudation—Strata identified by Superposition—By Fossils—Geological Record shown by Upturned Strata—General Result—Palæozoic and Primary Periods—Secondary—Tertiary—Time required—Coal Formation—Chalk—Elevations and Depressions of Land—Internal Heat of Earth—Earthquakes and Volcanoes—Changes of Fauna and Flora—Astronomical Time—Tides and the Moon—Sun's Radiation—Earth's Cooling—Geology and Astronomy—Bearings on Modern Thought.

CHAPTER III.

MATTER 51

Ether and Light—Colour and Heat—Matter and its Elements—Molecules and Atoms—Spectroscope—Uniformity of Matter throughout the Universe—Force and Motion—Conservation of Energy—Electricity, Magnetism, and Chemical Action—Dissipation of Heat—Birth and Death of Worlds.

CHAPTER IV.

LIFE 77

Essence of Life—Simplest Form, Protoplasm—Monera and Protista—Animal and Vegetable Life—Spontaneous Generation—Development of Species from Primitive Cells—Supernatural Theory—Zoological Provinces—Separate Creations—Law or Miracle—Darwinian Theory—Struggle for Life—Survival of the Fittest—Development and Design—The Hand—Proof required to establish Darwin's Theory as a Law—Species—Hybrids—Man subject to Law.

CHAPTER V.

ANTIQUITY OF MAN 105

Belief in Man's Recent Origin—Boucher de Perthes' Discoveries—Confirmed by Prestwich—Nature of Implements—Celts, Scrapers, and Flakes—Human Remains in River Drifts—Great Antiquity—Implements from Drift at Bournemouth—Bone Caves—Kent's Cavern—Victoria, Gower, and other Caves—Caves of France and Belgium—Ages of Cave Bear, Mammoth, and Reindeer—Artistic Race—Drawings of Mammoth, etc.—Human Types—Neanderthal, Cro-Magnon, Furfooz, etc.—Attempts to fix Dates—History—Bronze Age—Neolithic—Danish Kitchen-middens—Swiss Lake-Dwellings—Glacial Period—Traces of Ice—Causes of Glaciers—Croll's Theory—Gulf Stream—Dates of Glacial Period—Rise and Submergence of Land—Tertiary Man—Eocene Period—Miocene—Evidence for Pliocene and Miocene Man—Conclusions as to Antiquity.

CHAPTER VI.

MAN'S PLACE IN NATURE 166

Origin of Man from an Egg—Like other Mammals—Development of the Embryo—Backbone—Eye and other Organs of Sense—Fish, Reptile, and Mammalian Stages—Comparison with Apes and Monkeys—Germs of Human Faculties in Animals—The Dog—Insects—Helplessness of Human Infant—Instinct—Heredity and Evolution—The Missing Link—Races of Men—Leading Types and Varieties—Common Origin Distant — Language — How Formed — Grammar — Chinese, Aryan, Semitic, etc.—Conclusions from Language—Evolution and Antiquity — Religions of Savage Races — Ghosts and Spirits—Anthropomorphic Deities—Traces in Neolithic and Palæolithic Times—Development by Evolution—Primitive Arts — Tools and Weapons — Fire — Flint Implements — Progress from Palæolithic to Neolithic Times—Domestic Animals—Clothing—Ornaments—Conclusion, Man a Product of Evolution.

Part II.

MODERN THOUGHT.

CHAPTER VII.

MODERN THOUGHT 213

Lines from Tennyson—The Gospel of Modern Thought—Change exemplified by Carlyle, Renan, and George Eliot—Science becoming Universal—Attitude of Orthodox Writers—Origin of Evil—First Cause unknowable—New Philosophies and Religions—Herbert Spencer and Agnosticism—Comte and Positivism — Pessimism — Mormonism — Spiritualism — Dreams and Visions—Somnambulism—Mesmerism—Great Modern Thinkers—Carlyle—Hero-worship.

CHAPTER VIII.

MIRACLES 242

Origin of Belief in the Supernatural—Thunder—Belief in Miracles formerly Universal—St. Paul's Testimony—Now Incredible—Christian Miracles—Apparent Miracles—Real Miracles—Absurd Miracles—Worthy Miracles—The Resurrection and Ascension—Nature of Evidence required—Inspiration—Prophecy—Direct Evidence—St. Paul—The Gospels—What is Known of Them—The Synoptic Gospels—Resemblances and Differences—Their Origin—Papias—Gospel of St. John—Evidence rests on Matthew, Mark, and Luke—What each states—Compared with one another and with St. John—Hopelessly Contradictory—Miracle of the Ascension—Silence of Mark—Probable Early Date of Gospels—But not in their Present Form.

CHAPTER IX.

CHRISTIANITY WITHOUT MIRACLES 274

Practical and Theoretical Christianity—Example and Teaching of Christ—Christian Dogma—Moral Objections—Inconsistent with Facts—Must be accepted as Parables—Fall and Redemption—Old Creeds must be Transformed or Die—Mahometanism—Decay of Faith—Balance of Advantages—Religious Wars and Persecutions—Intolerance—Sacrifice—Prayer—Absence of Theology in Synoptic Gospels—Opposite Pole to Christianity—Courage and Self-reliance—Belief in God and a Future Life—Based mainly on Christianity—Science gives no Answer—Nor Metaphysics—So-called Intuitions—Development of Idea of God—Best Proof afforded by Christianity—Evolution is Transforming it—Reconciliation of Religion and Science.

CHAPTER X.

PRACTICAL LIFE 298

Conscience—Right is Right—Self-reverence—Courage—Respectability—Influence of Press—Respect for Women—Self-respect of Nations—Democracy and Imperialism—Self-knowledge—Conceit—Luck—Speculation—Money-making—Practical Aims of Life—Self-control—Conflict of Reason and Instinct—Temper—Manners—Good Habits in Youth—Success in Practical Life—Education—Stoicism—Conclusion.

SUPPLEMENTAL CHAPTER.

Gladstone's "Dawn of Creation" and "Proem to Genesis." Drummond's "Natural Law in the Spiritual World" 321

MODERN SCIENCE AND MODERN THOUGHT.

Part I.

MODERN SCIENCE.

MODERN SCIENCE & MODERN THOUGHT.

CHAPTER I.

SPACE.

Primitive Ideas—Natural Standards—Dimensions of the Earth—Of Sun and Solar System—Distance of Fixed Stars—Their Order and Size—Nebulæ and other Universes—The Telescope and the Infinitely Great—The Microscope and the Infinitely Small—Uniformity of Law—Law of Gravity—Acts through all Space—Double Stars, Comets, and Meteors—Has acted through all Time.

THE first ideas of space were naturally taken from the standard of man's own impressions. The inch, the foot, the cubit, were the lengths of portions of his own body, obviously adapted for measuring objects of comparatively small size with which he came in direct contact. The mile was the distance traversed in 1,000 double paces; the league the distance walked in an hour. The visible horizon suggested the idea that the earth was a flat, circular surface like a round table; and as experience showed that it extended beyond the limits of a single horizon, the conception was enlarged, and the size of the table increased so as to take in all the countries known to the geography of successive periods.

In like manner the sun, moon, and stars were taken

to be at the distance at which they appeared; that is, first of the visible horizon, and then of the larger circle to which it had been found necessary to expand it. It was never doubted that they really revolved, as they seemed to do, round this flat earth circle, dipping under it in the west at night, and reappearing in the east with the day. The conception of the universe, therefore, was of a flat, circular earth, surrounded by an ocean stream, in the centre of a crystal sphere which revolved in twenty-four hours round the earth, and in which the heavenly bodies were fixed as lights for man's use to distinguish days and seasons. The *maximum* idea of space was therefore determined by the size of the earth circle which was necessary to take in all the regions known at the time, with a little margin beyond for the ocean stream, and the space between it and the crystal vault, required to enable the latter to revolve freely. In the time of Homer, and the early Greek philosophers, this would probably require a maximum of space of from 5,000 to 10,000 miles. This dimension has been expanded by modern science into one of as many millions, or rather hundreds of millions, as there were formerly single miles, and there is no sign that the limit has been reached.

How has this wonderful result been arrived at, and how do we feel certain that it is true? Those who wish thoroughly to understand it must study standard works on Astronomy, but it may be possible to give some clear idea of the processes by which it has been arrived at, and of the cogency of the reasoning by which we are compelled to accept facts so contrary to the first impressions of our natural senses.

The fundamental principle upon which all measurements of space depend, which are beyond the actual application of human standards, is this: that distant objects change their bearings for a given change of base, more or less in proportion as they are less or more distant. Suppose I am on board a steamer sailing down the Thames, and I see two churches on the Essex coast directly opposite to me, or bearing due north, the first of which is one mile and the other ten miles distant. I sail one mile due east and again take the bearings. It is evident that the first church will now bear north-west, or have apparently moved through 45°, *i.e.*, one-eighth part of the circumference of a complete circle, assuming this circumference to be divided into 360 equal parts or degrees; while the more distant church will only have altered its bearing by a much less amount, easily determined by calculation, but which may be taken roughly at 5° instead of 45°.

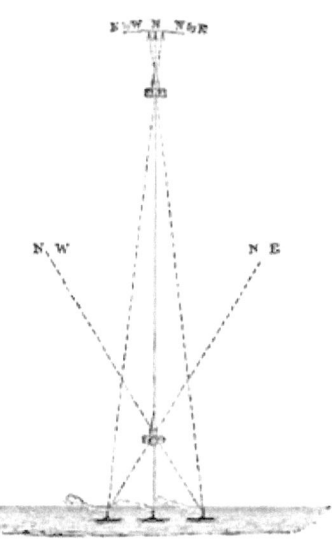

The branch of mathematics known as Trigonometry enables us in all cases, without exception, where we know the apparent displacement or change of bearing of a distant object produced by taking it from the opposite ends of a known base, to calculate the distance of that object with as much ease and certainty as if we were working a simple sum of rule of three. The first step is to know our base, and for this purpose it is essential to know the size and form of the earth on which we live. These are determined by very simple considerations.

If I walk a mile in a straight line, an object at a vast distance like a star will not change its apparent place perceptibly. But if I walk the same distance in a semi-circle, what was originally on my left hand will now be on my right, or will have changed its apparent place by 180°. If I walk my mile on the circumference of a circle of twice the size, I shall have traversed a quadrant or one-fourth part of it, and changed the bearing of the distant object exactly half as much, or 90°, and so on, according to the size of the circle, which may therefore be readily calculated from the length that must be travelled along it to shift the bearing of the remote object by a given amount, say of 1°.

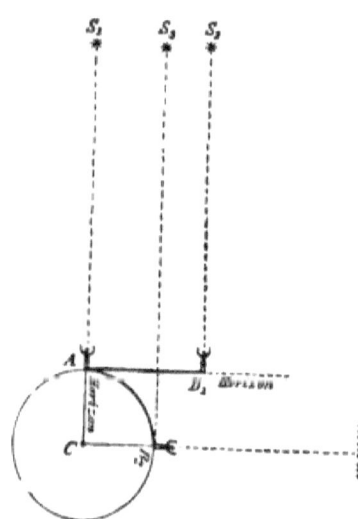

If, for instance, by travelling 65 miles from north to south we lower the apparent height of the Pole star 1°, it is mathematically certain that we have travelled this 65 miles, not along a flat surface, but along a circle which is 360 times 65, or, in round numbers, 24,000 miles in circumference and 8,000 miles in diameter. And if, whenever we travel the same distance on a meridian or line drawn on the circumference from north to south, we find the same displacement of 1°, we may be sure that our journey has been in a true circle, and that the form of the earth is a perfect sphere of these dimensions.

Now, this is very nearly what actually occurs when we apply methods of scientific accuracy to measure the earth. The true form of the earth is not exactly spherical, but slightly oval or flatter at the poles, being almost precisely the form it would have assumed if it had been a fluid mass rotating about a north and south axis. But it is very nearly spherical, the true polar diameter being 7,899 miles, and the true equatorial diameter 7,925 miles, so that for practical purposes we may say roughly that the earth is a spherical body, 24,000 miles round and 8,000 miles across.

This gives us a fresh standard from which to start in measuring greater distances. Precisely as we inferred the distance of the church from the steamer in our first illustration, we can infer the distance of the sun, from its displacement caused by observing it from two opposite ends of a base of known length on the earth's surface. This is the essential principle of all the calculations, though, when great accuracy is sought for, very refined methods of applying the principle are required, turning mainly on the extent to which the apparent occurrence of the same event—such as the transit of Venus over the sun's disc—is altered by observing it from different points at known distances from one another on the earth's surface. The result is to show that the sun's distance from the earth is, in round numbers, 93,000,000 miles. This is not an exact statement, for the earth's orbit is not an exact circle, but the sun and earth really revolve in ellipses about the common centre of gravity. The sun, however, is so much larger than the earth that this centre of gravity falls within the sun's surface, and, practically, the earth describes an ellipse about the sun, the 93,000,000 miles being the mean distance, and the eccentricity, or deviation from the exact circular orbit, being about one-sixtieth part of that mean distance. This distance, again, gives us the size of the sun, for it is easily calculated how large the sun must be to look as large as it does at a distance of 93,000,000 miles. The result is, that it is a sphere of about 880,000 miles in diameter. Its bulk, therefore, exceeds that of the earth in the proportion of 1,384,000 to 1. Its density, or the quantity of matter in it, may be calculated from the effect of its action on the earth under the law of gravity at the distance of 93,000,000 miles. It weighs as much as 354,936 earths.

The same method gives us the distance, size, and weight of the moon and planets; and it gives us a fresh standard or base from which to measure still greater distances. The distance of the earth from the sun being 93,000,000 miles, and its orbit an ellipse nearly circular, it follows that it is in mid-winter, in round numbers, 186,000,000 miles distant from the spot where it was at midsummer. What difference in the bearings of the fixed stars is caused by traversing this enormous base?

The answer is, in the immense majority of cases, no difference at all; *i.e.*, their distance is so vastly greater than 186,000,000 miles that a change of base to this extent makes no change perceptible to the most refined instruments in their bearings as seen from the earth. But the perfection of modern instruments is such, that a change of even one second, or $\frac{1}{3600}$th part of one degree, in the annual parallax, as it is called, of any fixed star, would certainly be detected.

This corresponds to a distance of 206,265 times

the length of the base of 186,000,000 miles, or of 20,000,000,000,000,000 miles, a distance which it would take light moving at the rate of 190,000 miles per second, three years and eighty-three days to traverse. There is only one star in the whole heavens, a bright star called Alpha, in the constellation of the Centaur, which is known to be as near as this. Its annual parallax is 0·976", or very nearly 1", and therefore its distance very nearly 20 millions of millions of miles. All the other stars, of which many millions are visible through powerful telescopes, are further off than this.

There are about eight other stars which have been supposed by astronomers to show some trace of an annual parallax of less than half a second, and therefore whose distances may be somewhere from twice to ten times as great as that of Alpha Centauri, and from the quantity of light sent to us from these distances, some approximation has been made to their intrinsic splendour as compared with our sun. That of Alpha Centauri is computed to be nearly $2\frac{1}{2}$ times, that of Sirius, the brightest star in the heavens, 393 times greater than that of the sun. These figures may or may not represent greater size or greater intensity of light, and they are only quoted to give some idea of the vastness of the scale of the universe, of which our solar system forms a minute part.

Nor does even this nearly fathom the depth of the abysses of space. Telescopes enable us to see a vast multitude of stars of varying size and brilliancy. It is computed by astronomers that there are at least one hundred millions of stars within the range of the telescopes used by Herschel for gauging the depth of space, and a thousand millions within the range of the great reflecting telescope of Lord Rosse. As many as eighteen different orders of magnitude have been counted, and the more the power of telescopes is increased the more stars are seen. Now, as there is no reason to suppose that this extreme variety of brilliancy arises from extreme difference of size of one star from another, it must be principally owing to difference of distance, so that a star of the eighteenth magnitude is presumably many times further off than any of the first magnitude, the distance of the nearest of which has been proved to be something certainly not less than 20,000,000,000,000 miles. In fact, these stellar distances are so great that in order to bring them at all within the range of human imagination we are obliged to apply another standard, that of the velocity of light. Light can be shown to travel at the rate of about 186 millions of miles in 16 minutes, for this is the difference of the time at which we see the same periodical occurrence, as for instance the eclipses of Jupiter's satellites, according as the earth happens to be at the point of its orbit nearest to Jupiter or at that farthest away. The velocity of light is therefore about 184,000 miles per second, a velocity which has been fully confirmed by direct experiments made on the earth's surface.

These enormous distances are reckoned, therefore, by the number of years which it would take light to come from them, travelling as it does at the rate of 184,000 miles a second. The nearest fixed star, Alpha Centauri, is seen by a ray which left it three years and eighty-three days ago, and has been travelling ever since at the rate of 184,000 miles per second. Sirius, the brightest of the fixed stars, if the determination of its annual parallax is correct, is six times further off, and is seen, not as it exists to-day, but as it existed nearly twenty years ago; and the light we now see from some of the stars of the eighteenth magnitude can hardly have left them less than 2,000 years ago.

Even this, however, is far from exhausting our conception of the magnitude of space. Beyond the stars which are near enough to be seen separately, powerful telescopes show a galaxy in which the united lustre of myriads of stars is only perceptible as a faint nebulous gleam. And in addition to stars the telescope shows us a number of nebulæ, or faint patches of light, sometimes globular, sometimes in wreaths, spiral wisps, and other fantastic shapes, scattered about the heavens. Some of these are resolved by powerful telescopes into clusters of stars inconceivably numerous and remote, which appear to be separate universes, like that of which our sun and fixed stars form one. Others again cannot be so resolved, and are shown by the spectroscope to be enormous masses of glowing gas, or cosmic matter, out of which other universes are in process of formation.

We are thus led, step by step, to enlarge our ideas of space from the primitive conception of miles and leagues, until the imagination fails to grasp the infinite vastness of the scale upon which the material universe is really constructed.

If the telescope takes us thus far beyond the standards of unaided sense in the direction of the infinitely great, the microscope, aided by calculations as to the nature of light, heat, electricity, and chemical action, takes us as far in the opposite direction of the infinitely small. The microscope enables us actually to see magnitudes of the order of $\frac{1}{100,000}$th of an inch as clearly as the naked eye can see those of $\frac{1}{10}$th. This introduces us into a new world, where we can see a whole universe of things both dead and alive of whose existence our forefathers had no suspicion. A glass of water is seen to swarm with life, and be the abode of bacteria, amœbæ, rotifers, and other minute creatures, which dart about, feed, digest, and propagate their species in this small world of their own, very much as jelly-fish and other humble organisms do in the larger seas. The air also is shown to be full of innumerable germs and spores floating in it, and ready to be deposited and spring into life, wherever they find a seed-bed fitted to receive them. Given a favourable soil in the human frame, and the invisible seeds of scarlet fever, cholera, and small-pox ripen into full crops, just as the germs of a fungus invade the potato crops of a whole district, and lead to Irish famines and the extermination of more than a million of human beings.

The microscope also enables us to see the very beginnings of life and watch its primitive element, protoplasm, in the form of a minute speck of jelly-like matter, through which pulsations are constantly passing, and we can watch the transformations by which an elementary cell of this substance splits up, multiplies, and by a continued process of development builds up with these cells all the diversified forms of vegetable and animal life.

But far as the microscope carries us down to dimensions vastly smaller than those of which the ordinary senses can take cognizance, the modern sciences of light, heat, and chemistry carry us as much farther downwards, as the telescope carries us upwards beyond the boundaries of our solar system into the expanses of stars and nebulæ. We are transported into a world of atoms, molecules, and light-waves, where the standard of measurement is no longer in feet or inches, or even in one-hundred-thousandth part of an inch, but in millionths of millimetres, i.e., in $\frac{1}{25,000,000,000}$th of an inch. The dimensions are such that, as we shall see when we come to deal with matter, if the drop of water in which the microscope shows us living animalcula were magnified to the size of the earth, the atoms of which it is composed would appear of a size intermediate between that of a rifle-bullet and a cricket-ball.

This, then, is Nature's scale of space, from millionths of a millimetre up to millions of millions of miles. Throughout the whole of this enormous range of space the laws of Nature prevail.

Matter attracts matter by the same law of gravity in the case of double stars revolving about each other at a distance at which a base of 180,000,000 miles has long since become a vanishing point, and in the case of atoms which form the substance of a gas, as in that of an apple falling from a tree at the earth's surface. Comets, darting off into the remote regions of space, return after long periods, in obedience to the same law. Clouds of meteoric dust revolve in fixed orbits, determined by the law of gravity as surely as the moon revolves round the earth, and the earth round the sun.

This is a conclusion of such fundamental importance that it is desirable to give the uninitiated reader some clear idea of what it means, and how it is arrived at. Newton's great discovery, the law of gravity, is this—that all matter acting in the mass attracts other matter directly as the amount of attracting matter, and inversely as the square of the distance. That is, 2 or 2,000,000 tons attract with twice the force of 1 or 1,000,000 tons at the same distance, but with only one-fourth of the same force at double, and one-ninth at triple the distance.

How is this law proved? This will be best answered by explaining how it was discovered. The force of gravity, or attraction of the earth on bodies at the earth's surface, is a known quantity. The whole matter in a spherical body attracts exactly as if it were all collected at the centre. The force of gravity at the earth's surface is, therefore, that of the earth's mass exerted at a distance of about 4,000 miles, and this can be easily measured by observing the space fallen through, and the velocity acquired, by a falling body in a given time, such as 1″.

Does the same force act at the distance of the moon, or 207,200 miles? This was the question Newton asked himself, and the answer was got at in the following way. If we swing a stone in a sling round our head, it describes a circle as long as we keep the string tight, and its pull inwards just balances the pull of the stone to fly outwards, *i.e.*, to use scientific language, as long as the centripetal just balances the centrifugal force. But if we let go the string the stone darts off in the direction in which, and with the velocity with which, it was moving when the centripetal force ceased to act.

The moon is such a sling-stone revolving about the earth. At each instant it is moving in the direction of a tangent to its orbit, and would move on in a straight line along this tangent if it were not deflected from it by some other force. That is, if the moon were now at M_1, it would, after a given interval of time, be at M_2 if no force had acted on it. But in point of fact it is **not** at M_2 but at M_3. Therefore it has been pulled down from M_2 to M_3, or, if you like, fallen through the space $M_2 M_3$ in the time in which it would have travelled over $M_1 M_2$ with its velocity at M_1. How does this space correspond with the space through which a heavy body would have fallen in the same time at the earth's surface? It corresponds exactly, assuming the law of gravity to be, that it decreases with the square of the distance.

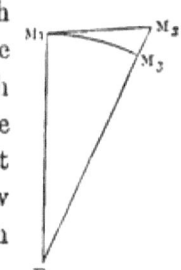

This may be taken as the first approximation, but the more accurate and universal proofs of the law are derived from mathematical calculations of what the nature of the attractions must be, in the case of the sun, earth, moon, and planets, to make them describe such elliptic orbits and observe such laws, as from Kepler's observations we know actually to be the case. The answer here again is the law of gravity, and no other possible law, and this is confirmed in practice by the fact that we are able, by calculations based on it, to satisfy the requisite of safe prophecy—that of knowing beforehand, and to predict eclipses, comets, transits, and occultations, and generally to compile Nautical Almanacs, by which ships know their whereabouts in pathless oceans.

This, then, affords us a first firm standing-point in any speculations as to the nature of the universe. One great law, at any rate, is universal throughout all space, and, as we shall see later, suns, stars, and nebulæ are composed of the same matter as the earth and its inhabitants.

In like manner comets and meteors, though presenting in other respects phenomena not yet fully understood, are proved to obey the same laws and to consist of the same matter. Comets are bodies which revolve round the sun, and are attracted by it and by the planets, in obedience to the ordinary law of gravity, though their density is so slight, that although often of enormous volume, they produce no perceptible effect on the planets, even when entangled amidst the satellites of a planet, as Lascelles' comet was among those of Jupiter.

Their dimensions may be judged of when it is stated that the comet of 1811 had a tail 120 millions of miles in length and 15 millions of miles in diameter at the widest part, while the diameter of the nucleus was about 127,000 miles, or more than ten times that of the earth. In order that bodies of this magnitude, passing near the earth, should not affect its motion or change the length of the year by even a single second, their actual substance must be inconceivably rare. If the tail, for instance, of the comet of 1843 had consisted of the lightest substance known to us, hydrogen gas, its mass would have exceeded that of the sun, and every planet would have been dragged from its orbit. As Proctor says, therefore: "A jar-full of air would probably have outweighed hundreds of cubic miles of that vast appendage which blazed across the skies to the terror of the ignorant and superstitious."

The extreme tenuity of a comet's mass is also proved by the phenomenon of the tail, which, as the comet approaches the sun, is thrown out sometimes to a length of 90 millions of miles in a few hours. And what is remarkable, this tail is thrown out against the force of gravity by some repulsive force, probably electrical, so that it always points away from the sun.

Thus a comet which approaches the sun with a tail behind it, will, after passing its perihelion, recede from the sun with its tail before it, and this although the tail may be of the length of 200 millions of miles as in the comet of 1843. In the course of a few hours, therefore, this enormous tail has been absorbed and a new one started out in an opposite direction. And yet, thin as the matter of comets must be, it obeys the common law of gravity, and whether the comet revolves in an orbit within that of the outer planets, or shoots off into the abysses of space and returns only after hundreds of years, its path is, at each instant, regulated by the same force as that which causes an apple to fall to the ground; and its matter, however attenuated, is ordinary matter, and does not consist of any unknown elements. The spectroscope shows that comets shine partly by reflected sunlight and partly by light of their own, the latter part being gaseous, and this gas, in most comets, contains carbon, hydrogen, and nitrogen, possibly also oxygen, in the form of hydrocarbons or marsh gas, cyanogen and possibly oxygen compounds of carbon. One comet has recently given the line of sodium, and the presence of iron is strongly suspected.

As regards meteors, which include shooting stars and aërolites, it has been long known, from actual masses which have fallen on the earth, that they are composed of terrestrial matter, principally of iron, which has been partially fused by the heat engendered by the friction of the rapid passage through the air. The recurrence of brilliant displays at regular intervals, as for instance those of August and November, when the whole sky often seems alive with shooting stars, had also been noticed; but it was reserved for recent times to prove

c

that these meteor streams are really composed of small planetary bodies revolving round the sun in fixed orbits by the force of gravity, and that their display, as seen by us, arises from the earth in its revolution round the sun happening to intersect some of these meteoric orbits, and the friction of our atmosphere setting fire to and consuming the smaller meteors which appear as shooting stars. This shows the enormous number of meteors by which space must be tenanted. It is proved that the earth encounters more than a hundred meteor systems, but the chance of any one ring or system being intersected by the earth is extremely small, as the earth is such a minute speck in the whole sun-surrounding space of the solar system. On a scale on which the earth's orbit was represented by a circle of 10 feet diameter, the earth itself would be only about $\frac{1}{100}$th of an inch in diameter, so that if, as astronomers say, the earth encounters about a hundred meteor systems in the course of its annual revolution, space must swarm with an innumerable number of these minute bodies all revolving round the sun by the force of gravity.

Has this law of gravity been uniform through all time as it undoubtedly is through all space? We have every reason to believe so. The law of gravity, which is the foundation of most of what we call the natural laws of geological action, has certainly prevailed, as will be shown later, through the enormous periods of geological time, and far beyond this we can discern it operating in those astronomical changes by which cosmic matter has been condensed into nebulæ, nebulæ into suns throwing off planets, and planets throwing off satellites, as they cooled and contracted. We cannot speak with quite the same certainty of infinite time as we can of infinite space, for we have no telescopes to gauge the abysses of time, and no certain standards, like those of the known dimensions of our solar system, to apply to periods too vast for the imagination.

But we can say this with certainty, that the present law of gravity must have prevailed when the outermost planet of our system, Neptune, was condensed into a separate body and began revolving in its present orbit, and that it has continued to act ever since; while, as a matter of probability, it is as nearly certain as anything can be, that the law by which the apple falls to the ground is an original law of matter, and has existed as long as matter has existed.

It certainly extends through all space. Double stars at a distance exceeding 20 millions of millions of miles revolve round their common centre of gravity by this law. Atoms and molecules almost infinitely smaller than millionths of millimetres derive from it their specific weights with as much certainty as if they were pounds or hundredweights.

What space and matter really may be, we do not know, and if we attempt to reason about their essence and origin, or quit the region of science based on fact, we get into the misty realms of metaphysics, where, like Milton's fallen angels, we

<blockquote>Find no end in wandering mazes lost.</blockquote>

But this we do know of a certainty, that be matter and space what they may, they are subject to this one, uniform, all-pervading law; and attract, have always attracted, and will always attract, directly as the mass of the attracting matter and inversely as the square of the distance in space at which the attraction acts.

CHAPTER II.

TIME.

Evidence of Geology—Stratification—Denudation—Strata identified by Superposition—By Fossils—Geological Record shown by Upturned Strata—General Result—Palæozoic and Primary Periods—Secondary—Tertiary—Time required—Coal Formation—Chalk—Elevations and Depressions of Land—Internal Heat of Earth—Earthquakes and Volcanoes—Changes of Fauna and Flora—Astronomical Time—Tides and the Moon—Sun's Radiation—Earth's Cooling—Geology and Astronomy—Bearings on Modern Thought.

GEOLOGY has done for time what astronomy has for space—it has expanded the limited ideas derived from natural impression and early tradition, into those of an almost infinite duration. This result is so important that it is desirable that all educated persons, without being professed geologists, should have some clear idea of the nature of the conclusions and of the evidences on which they rest.

This I will endeavour to give.

When we come to examine the structure of the earth —or rather of the outer crust of the earth which we inhabit—with the care and precision of scientific methods, we find that it is not of uniform composition, but consists mainly of distinct layers, or strata, lying one over the other. This is true not only of the larger beds, or distinct formations, but of the details of each formation, many of which are built up as regularly as the layers of the Great Pyramid, while others are made up of layers no thicker than the leaves of a book.

Now consider what this fact of stratification implies. In the first place it implies deposit from water, for there is no other agency by which materials can be sorted out and thrown down in horizontal layers, while this agency is now doing the same thing every day and all over the world. The Rhone flows into the Lake of Geneva a turbid stream, and flows out of it as clear as crystal. All the matter it brings in is deposited at the bottom of the lake, and in course of time will fill it up. This deposit varies with every alternation of flood and drought; the river depositing sometimes boulders and coarse gravel, sometimes shingle, sand, or fine mud, and carrying this material sometimes to a greater and sometimes to a less distance, according to the velocity of the stream.

Ages hence, when the lake has been converted into dry land, it will be as certain, whenever a pit is dug or a well sunk in it, that it was the work of a river flowing into a lake, as it is to-day, when we can see them at work.

And what is true of the Rhone and the Lake of Geneva, is true on a larger scale of the Ganges, the Mississippi, and of every sea or ocean, with every river or torrent pouring into it.

Again, the sea is perpetually wearing away the coasts of all lands, and, where the cliffs are soft and the tides and currents strong, at a very rapid rate. The materials swallowed up are rolled as shingle, ground into sand, or floated as fine mud, and all finally assorted and laid down at the bottom of the sea, not in a confused heap, but in regular succession. On some of them, shell-fish and other marine creatures live and die for generations, and their remains are covered over by fresh sands or clays, and preserved for future geologists. All this is going on now, and when we examine the rocks we find that precisely the same sort of thing has been going on from the newest to the oldest strata. With the exception of a comparatively small amount of igneous rock, which has boiled up from deep sources of molten matter, and been poured out in sheets of lava, or masses of trap, porphyry, and granite, according to the amount of pressure it has undergone and the time it has taken to cool and crystallise, all the earth's surface may be said to consist of stratified matter, showing clear signs of having been deposited from water. Some of the oldest rocks, such as gneiss, may be a little doubtful, as they have clearly been subjected to great heat under great pressure, until they became plastic enough to crystallise as they cooled, and thus destroy any fossils embedded in them and obliterate most of the ordinary signs of stratification. But the opinion of the best geologists is that they were originally stratified, and have become what is called "metamorphic," or changed by heat and pressure into the semblance of igneous rocks. But even if these are not included, enough remains to justify the general assertion that the outer crust of the earth, as known to us, is made up mainly of stratified materials which have been deposited from water.

Now this implies another most important fact, viz., that there must have been waste or denudation of existing land corresponding to the deposit of stratified materials under water. Water cannot generate these materials, and every square mile of such strata, say 10 feet thick, implies the removal of 10 feet from a square mile of land surface by rains and rivers, or of an equivalent amount of cubical content in some other way, as by the erosion of a coast line. This is a very important consideration when we come to estimate the time required for the formation of such a thickness of stratified beds as we find existing. There must have been a fundamental crystalline rock as the earth cooled down from a fluid state and acquired a solid crust, and this rock must have been worn down by primeval seas and rivers as the progressive cooling admitted of the condensation of aqueous vapour into water. The waste of this primitive crust must have been deposited in strata at the bottom of those seas in thick masses, covering the original rock, and these again must have been partly crystallised by heat and pressure, and over and over again upheaved and submerged, and themselves worn down by fresh erosion, forming fresh deposits which underwent a repetition of the same process.

A third important inference from the fact of stratification is that all strata must have been originally deposited horizontally, or very nearly so, and in such order that the lowest is the oldest.

Suppose we fill a jar with water, and put some white sand into it, and when that has subsided to the bottom and the water is clear, some yellow sand, and again some red sand, it is clear that we shall have at the bottom of the jar three horizontal deposits or strata, one white, one yellow, and one red, and that by no conceivable means can the order in which they were deposited have been other than first white, secondly yellow, and lastly red. This law, therefore, is invariable, that wherever it is possible to trace a series of strata lying one above the other, the lowest is the oldest, and the highest the youngest in point of time.

If, therefore, all the great formations, from the old Laurentian up to the newest Tertiary, had been deposited uniformly all over the world, and had remained undisturbed, and we could have seen them in one vertical section in a cliff twenty-five miles high—for that is about their total known thickness—we should have been able without further difficulty to determine their order of succession and respective magnitudes.

But this is plainly impossible, for the deposits going on at any one time are of very different character. For instance, we have at present the Globigerina ooze gradually filling the depths of the Atlantic with a deposit resembling chalk; the Gulfs of Bengal and Mexico silting up with fine clay from river deposits; vast tracts in the Pacific, Indian Ocean, and Red Sea, covered with coral and the *débris* of coral-reefs. How could these, if upheaved into dry land and explored by future geologists, be identified as having been formed contemporaneously?

Suppose that coins of Victoria had been dropped in each of them, the geologist who discovered these coins would have no difficulty in concluding that the strata in which they were found were all formed in the nineteenth century. The petrified shells and other remains found in geological strata are such coins. Every great formation has had its own characteristic fauna and flora, or aggregate of animal and vegetable life, varying slowly from one geological age to another, and linked to the past and future by some persistent types and forms, but still with such a preponderance of characteristic fossils as to enable us to assign the rocks in which they occur to their proper place in the volume of the geological record. Innumerable observations have shown that we can rely, with absolute confidence, on the fossils embedded in the different strata of the earth's crust as tests of the period to which they belong, however different the strata may be in mineral composition.

The next question is how we can ascertain the thickness and order of succession of these strata. We have seen that all stratified rocks were originally deposited from water and therefore horizontally. Had they remained so, in the first place the process of forming stratified rocks must long ago have come to an end, for all the land surface must have been worn down to the sea level, and with no more land to be denuded, deposition must have ceased at an early period of the earth's history. And, in the second place, we could have known nothing more of the earth's crust than we saw on the surface, and in the shallow pits and borings we could sink below it. But earthquakes and volcanoes, and the various fractures and pressures due to subterranean heat and secular contraction and cooling, have been at work counteracting the effects of denudation, and causing elevations and depressions by which the inequalities of the earth's surface have been renewed, the balance between sea and land maintained, and strata, originally horizontal at the bottom of the ocean, upheaved until sea-shells are found at the top of high mountains, and we can walk for miles over their upturned edges.

Any one who wishes to understand how geologists have been able to measure such a thickness of the earth's crust, has only to take a book open at page 1 and lay it flat before him. He can see nothing but that one page; but if he turns up the pages on the right-hand side of the book until their edges become horizontal, he can pass over them and count perhaps 500 pages in the space of a couple of inches.

This is precisely what geologists have been able to do at various points of the earth's surface where the upturned edges of the pages of its history are exposed, and they come out, one behind the other, in the due succession in which they were written by Nature. For instance, in travelling from east to west in England we pass continually from newer to older formations—Chalk comes in from below Tertiary; Oolite and Lias from below Chalk; then Permian or New Red Sandstone; Carboniferous, including the Coal measures; Devonian or Old Red Sandstone; Silurian, Cambrian, and in the extreme north-west of Scotland and the Hebrides, oldest of all the Laurentian.

There are some omissions and interpolations, but, in a general way, it may be said that within the bounds of the British Empire we have such a view of Nature's volume as would be got, in the case I have supposed, by travelling over its upturned edges from page 1 to page 500. And if each of the great formations be taken as a separate chapter, each chapter will be found to be made up of a number of pages, each with its own letterpress and illustrations, though connected with the pages before and after it by the thread of the continuous common subject of their proper chapter; as the chapters again are connected by the continuous common subject-matter of the complete volume. It must not be supposed that the volume is anything like perfect. We have to piece it together from fragments found in the limited number of countries which have thus far been scientifically explored, and which do not constitute more than a small part of the earth's surface. We know nothing of what is below the oceans which cover three-fourths of that surface, and there are great gaps in the record during times when portions of the surface were dry land, and consequently no deposit of strata or preservation of fossils was possible. Still a great deal has been accomplished, and the general result, as given by common consent of the best geologists, is as follows :

The total thickness of known strata is about 130,000 feet or twenty-five miles, or the $\frac{1}{100}$th part of the distance from the earth's surface to its centre. Of this, about 30,000 feet belong to the Laurentian, which is the oldest known stratified deposit; 18,000 to the Cambrian, and 22,000 to the Silurian. These form together what is known as the Primary or Palæozoic Epoch.

In the lowest, the Laurentian, the only faint trace of life discovered is that of the Eozoon Canadense, which is considered to be an undoubted petrifaction of a foraminiferous living organism with a chambered shell.

It must be remembered, however, that these earliest formations have been so changed by slow crystallisation under great heat and pressure that all fossils and nearly all traces of stratification must have been obliterated.

In the Cambrian and Lower Silurian traces of life become more frequent, especially of low forms of seaweeds, and in the Upper Silurian we find an abundance of life, consisting of crustacea, shell-fish, and a few true fish in the upper strata. Some of these shells, as the Lingula, have continued without much change up to the present time; and on the whole we find ourselves in the

Silurian period, if not earlier, in presence of a state of things in which substantially present causes operated and present conditions were in force. Rains fell, winds blew, rivers ran, waves eroded cliffs, shell-fish lived and died, and crabs and sand-worms crawled about on shores left dry by each tide, very much as is the case at present.

The next great division, which got the name of Primary before the existence of fossils was known in the older or Palæozoic division, comprises the Devonian or Old Red Sandstone; the Carboniferous, which includes the coal; and the Permian or New Red Sandstone. The average thickness of these three systems taken together is about 42,000 feet. It may be called the era of Fern Forests and of Fish, the former being the principal source of our supplies of coal, and the latter being extremely abundant within the Devonian and Permian formations.

The third great division is formed by the Secondary group, which includes the Triassic, the Jura, and the Cretaceous or Chalk systems, and has an average thickness of about 15,000 feet. This epoch is emphatically the age of Reptiles as the preceding one was that of Fish, and the prevailing vegetation is no longer one of ferns and mosses, but of Gymnosperms, or plants having naked seeds, the most important class of which is that of the Coniferæ or Pine tribe. During this period the Plesiosauri, Ichthyosauri, and other gigantic sea-dragons abounded in the oceans; colossal land-dragons, such as the Dinosauri, occupied the continents, and Pterodactyls, a remarkable form of carnivorous flying lizards, ruled the air. Swarms of other reptiles, nearly related to the present lizards, crocodiles, and turtles, abounded both in the sea and land. A few traces of mammals and birds show that these orders had then come into existence, just as a few traces of reptiles are found in the Primary and of fish in the Palæozoic strata, but the few mammalian remains found are of small animals of the marsupial or lowest type, and the birds are of a transition type between reptiles and true birds. This epoch concludes with the Chalk formation, which is one of deep-sea deposit, where no trace of terrestrial life can be expected.

Above this comes the Tertiary epoch, when the present order, both of vegetable and animal life, is fairly inaugurated; mammals predominate over other forms of vertebrate animals; existing orders and species begin to appear and increase rapidly; and vegetation consists mainly of Angiosperms, or plants with covered seeds, as in our present forests. The total thickness of these strata, from the lowest, or Eocene, to the end of the uppermost, or Pliocene, is about 3,000 feet. Above this comes the Quaternary, or recent period, which comprises the superficial strata of modern formation, and is characterised by the undoubted existence of man and of animal species, which either now exist or have become extinct in quite recent geological times.

The details of this and of the Tertiary Epoch will be more fully considered when we come to treat of the antiquity of man, with which they are closely connected. But for the present object, which is that of ascertaining some standard of time for the immense series of ages proved by geology to have elapsed since the earth assumed its present condition, became subject to existing laws and fitted to be the abode of life, it will be sufficient to refer to the older strata.

The best idea of the enormous intervals of time required for geological changes will be derived from the coal measures. These consist of part only of one geological formation known as the Carboniferous. They are made up of sheets, or seams, of condensed vegetable matter, varying in thickness from less than an inch to as much as thirty feet, and lying one above another, separated by beds of rock of various composition. As a rule, every seam of coal rests upon a bed of clay, known as the "under-clay," and is covered by a bed of sandstone or shale. These alternations of clay, coal, and rock, are often repeated a great many times, and in some sections in South Wales and Nova Scotia, there are as many as eighty or a hundred seams of coal, each with its own under-clay below and sandstone or shale above. Some of the coal seams are as much as thirty feet thick, and the total thickness of the coal measures is, in some cases, as much as 14,000 feet.

Now consider what these facts mean. Every under-clay was clearly once a surface soil on which the forest vegetation grew, whose accumulated *débris* forms the overlying seam of coal. The under-clays are full of the fibres of roots, and the stools of trees which once grew on them, are constantly found *in situ*, with their roots attached just as they stood when the tree fell, and added to the accumulation of vegetable matter, which in modern times forms peat, and in more ancient days, under different conditions of heat and pressure, took the more consolidated form of coal.

When these vegetable remains are examined with the aid of the microscope it is found that these ancient forests consisted mainly of trees like gigantic club-mosses, mares'-tails, and tree ferns, with a few resembling yews and firs. But in many cases the bulk of the coal is composed of the spores and seeds of these ferns and club-mosses, which were ripened and shed every year, and gradually accumulated into a vegetable mould, just as fallen leaves, beech-mast, and other *débris*, gradually form a soil in our existing forests.

The time required must have been very great to accumulate vegetable matter, principally composed of fine spore dust, to a depth sufficient under great compression to give even a foot of solid coal. Dr. Dawson, who has devoted great attention to the coal-fields of America, says: "We may safely assert that every foot of thickness of pure bituminous coal implies the quiet growth and fall of at least fifty generations of Sigillaria, and therefore an undisturbed condition of forest growth, enduring through many centuries." But this is only the first step in the measure of the time required for the formation of the coal measures. Each seam of coal is, as we have seen, covered by a bed of sand or shale, *i e.*, of water-borne materials. How can this be accounted for? Evidently in one way only—that the land surface in which the forest grew subsided gradually until it became first a marsh, and then a lagoon or shallow estuary, which silted up by degrees with deposits of sand or mud, and, finally, was upraised until its surface became dry land, in which a second forest grew, whose *débris* formed a second coal seam. And so on, over and over again, until the whole series of coal measures had been accumulated, when this alternation of slight submergences and slight rises came to an end, and some more decided movement of the earth's surface in the locality brought on a different state of things. This is in fact exactly what we see taking place on a smaller scale in recent times in such deposits as those of the delta of the Mississippi, where a well sunk at New Orleans passes through a succession of cypress swamps and forest growths, exactly like those now growing on the surface, which are piled one above the other, and separated by deposits of river silt, showing a long alternation of periods of rest when forests grew, followed by periods of subsidence when they were flooded and their remains were embedded in silt.

Starting on Dr. Dawson's assumption that one foot of coal represents fifty generations of coal plants, and that each generation of coal plants took ten years to come to maturity, an assumption which is certainly very moderate, and taking the actually measured thickness of the coal measures in some localities at 12,000 feet, Professor Huxley calculates that the time represented by the Coal formation alone would be six millions of years. Such a figure is, of course, only a rough approximation, but it is sufficient to show that when we come to deal with geological time, the standard by which we must measure is one of which the unit is a million of years.

This standard is confirmed by a variety of other considerations. Take the case of the Chalk formation.

Chalk is almost entirely composed of the microscopic shells of minute organisms, such as now float in the upper strata of our great oceans, and by their subsidence, in the form of an impalpable shell-dust, accumulate what is called the "Globigerina ooze," which is brought up by soundings in the Atlantic and Pacific from great depths. In fact, we may say that a chalk formation is now going on in the depths of existing oceans, and conversely that the old chalk, which now forms hills and elevated downs, was certainly deposited at the bottom of similar deep oceans of the Cretaceous period. The rate of deposit must have been extremely slow, certainly much slower than that of the deposit of the much grosser matter brought down by the Nile in its annual inundations, the growth of which has been estimated from actual measurement at about three inches per century. If one inch per century were the rate of accumulation of this microscopic shell-dust, subsiding slowly to depths of two or three miles over areas as large as Europe, it would take 1,200 years to form a foot of chalk, and 1,200,000 years to form 1,000 feet. Now there are places where the thickness of the Cretaceous formation, exposed by the edges of its upturned strata, exceeds 5,000 feet, so that this gives an approximation very similar to that furnished by the coal measures.

We have thus, on a rough approximation, a *minimum* period of about 6,000,000 years for the accumulation of a single member of one of the separate formations into which the total 130,000 feet of measured strata are subdivided. But this takes no account of the long periods during which no accumulation took place at the localities in question, and of the long pauses which must have ensued between each movement of elevation and submergence, and especially between the disappearance of an old and appearance of an almost entirely new epoch, with different forms of animal and vegetable life. We may be certain also that we are far from knowing the total thickness of strata which will be disclosed when the whole surface of the earth comes to be explored. All we can say is that we have fragmentary pages left in the geological record for, at the very least, 100 millions of years, and that probably the lost pages are quite as numerous as those of which we have an imperfect knowledge.

Sir Charles Lyell, the highest authority on the subject, is inclined to estimate the *minimum* of geological time at 200 millions of years, and few geologists will say that his estimate appears excessive.

Another test of the vast duration of geological time is afforded by the oscillations of the earth's surface. At first sight we are apt to consider the earth as the stable and the sea as the unstable element. But in reality it is exactly the reverse. Land has been perpetually rising and falling while the level of the sea has remained the same. This is easily proved by the presence of sea-shells and other marine remains in strata which now form high mountains. In the case of chalk, for instance, there must have been in England a change of relative level of sea and land of more than two miles of vertical height, between the original formation of the chalk at the bottom of a deep ocean and its present position in the North and South Downs. In other cases the change of level is even more conspicuous. The Nummulite limestone, which is formed like chalk from an accumulation of the minute shells of low organisms floating in the oceans of the early Tertiary period, is found in mountain masses, and has been elevated to a height of 10,000 feet and more in the Alps and Himalayas.

On a smaller scale, and in more recent times, raised beaches with existing shells and lines of cliffs and caves, are found at various heights above the existing sea-level of many of the coasts of Britain, Scandinavia, Italy, South America, and other countries.

Now the first question is, were these changes caused by the land rising or by the sea falling? The answer is, by the land rising. Had they been caused by the sea standing at a higher level it must have stood everywhere at this level, at any rate in the same hemisphere and anywhere near the same latitude. But there are large tracts of land which have never been submerged since remote geological periods; and in recent times there is conclusive evidence that the changes of level of sea and land have been partial and not general. Thus in the well-known instance of the columns of the ruined temple of Serapis at Pozzuoli in the Bay of Naples, which forms the illustration on the title-page of Lyell's "Principles of Geology," there can be no doubt that since the temple was built, either the sea must have risen and since fallen, or the land sunk and since risen, at least twenty feet since the temple was built less than 2,000 years ago, for up to this height the marble columns are riddled by borings of marine shells, whose valves are still to be seen in the holes they excavated. But an elevation of the level of the Mediterranean of twenty feet would have submerged a great part of Egypt, and other low-lying lands on the borders of that sea, where we know that no such irruptions of salt water have taken place within historical, or even within recent geological times.

The conclusion is therefore certain, that the land at this particular spot must have sunk twenty feet, and again risen as much, so as to bring back the floor of the temple to its present position, which stood one hundred years ago just above the sea-level, and that so gradually as not to throw down the three columns which are still standing. A slow subsidence has since set in and is now going on, so that the floor is now two or three feet below the sea-level.

Similar proofs may be multiplied to any extent. Along the coasts of the British Islands we find, in some places submarine forests showing subsidence, in others raised beaches showing elevation, but they are not continuous at the same level. Along the east coast of Scotland there is a remarkable raised beach at a level of about twenty-four feet above the present one, showing in many places lines of cliff, sea-worn caves, and outlying stacks and skerries, exactly like those of the present coast, though with green fields or sandy links at their base, instead of the waves of the German Ocean. But as we go north this inland cliff gets lower and gradually dies out, and when we get into the extreme north, among the Orkney and Shetland Islands, there are no signs of raised beaches, and everything points towards the recent period having been one of subsidence.

Again, in Sweden, where marks were cut in rocks in sheltered situations on the tideless Baltic more than a century ago, so as to test the question of an alleged elevation of the land, it has been clearly shown that, in the extreme north of Sweden, the marks have risen nearly seven feet, while in the central portion of the country they have neither risen nor fallen, and in the southern province of Scania they have fallen.

This would be clearly impossible if the sea and not the land had been the unstable element, and apparent elevations and depressions had been due to a general fall or rise in the level of all the seas of the northern hemisphere.

In fact, the more we study geology the more we are

impressed with the fact that the normal state of the earth is, and has always been, one of incessant changes. Water, raised by evaporation from the seas, falls as rain or snow on land, wastes it away and carries it down from higher to lower levels, to be ultimately deposited at the bottom of the sea. This goes on constantly, and if there were no compensating action, as the seas cover a much larger area than the lands, all land would ultimately disappear, and one universal ocean cover the globe. But inward heat supplies the compensating action, and new lands rise and new mountain chains are upheaved to supply the place of those which disappear.

This inward heat of the earth is not a mere theory, but an ascertained fact; for as we descend from the surface in deep mines or borings, we find the temperature actually does increase at a rate which varies somewhat in different localities, but which averages about 1° Fahrenheit for every 60 feet of depth. At this rate of increase water would boil at a depth of 10,000 feet, and iron and all other metals be melted before we reached 100,000 feet. What actually occurs at great depths we do not know with any certainty, for we are not sufficiently acquainted with the laws under which matter may behave when under enormous heat combined with enormous pressure. But we do know from volcanoes and earthquakes that masses of molten rocks and of imprisoned gases exist in certain localities, at depths below the surface which, although large compared with our deepest pits, are almost infinitesimally small compared with the total depth of 4,000 miles from that surface to the earth's centre.

This much is clear, that, in order to account for observed facts, we must consider the extreme outer crust, or surface of the earth as known to us, as resting on something which is liable to expand and contract slowly with variations of heat, and occasionally, when the tension becomes great, to give violent shocks to the outer crust, sending earthquake waves through it, and to send up gases and molten lava through volcanoes, along lines of fissure, and at points of least resistance. It is clear, also, that these movements are not uniform, but that one part of the earth's surface may be rising while another is sinking, and portions of it may be slowly tilting over, so that as one end sinks the other rises.

The best comparison that can be made is to a sheet of ice which has been much skated over and cracked in numerous directions, so as to have become a sort of mosaic of ice fragments, which, when a thaw sets in and the ice gets sloppy, rise and fall with slightly different motions as a skater, gliding over them, varies the pressure, and occasionally give a crack and let water rise through from below in the line of fissure. The difficulty will not seem so great if we consider that the rocks which form the earth's crust are for the most part elastic, and that an amount of elevation which seems large in itself does not necessarily imply a very steep gradient. Thus, if the elevation which towards the close of the Glacial period carried a bed of existing sea-shells of Arctic type to the top of the hill, Moel Tryfen, in North Wales, which is 1,200 feet high, were, say one of 1,500 feet, this would be given by a gradient of 15 feet a mile, or 1 in 333 for 100 miles. Such a gradient would not be perceptible to the eye, and would certainly not be sufficient to cause any tension likely to rupture rocks or disturb strata.

Such movements are as a rule extremely slow. In volcanic regions there are occasionally shocks which raise extensive regions a few feet at a blow, and partial elevations and subsidences which throw up cones of lava and cinders, or let mountains down into chasms, in a single explosion. The most noted of these are the instances of Monte Nuovo, near Naples, 800 feet high, and Jorullo, in Mexico, thrown up in one eruption, and the disappearance the other day of a mountain 2,000 feet high in the Straits of Sunda during an earthquake. The largest rise recorded of an extensive area from the shock of an earthquake, is that which occurred in South America in 1835, when a range of coast of 500 miles from Copiapo to Chiloe was permanently raised five or six feet by a single shock, as was shown by the beds of dead mussels and other shells which had been hoisted up in some places as much as ten feet. It is probable that the great chain of the Andes, whose highest summits reach 27,000 feet, has been raised in a great measure by a succession of similar shocks.

But for the most part these movements, whether of elevation or depression, go on so slowly and quietly that they escape observation. Scandinavia is apparently now rising and Greenland sinking, but most countries have remained appreciably steady, or nearly so, during the historical period. St. Michael's Mount, in Cornwall, is still connected with the mainland by a spit, dry at ebb tide and covered at flood, as it was more than 2,000 years ago when the old Britons carted their tin across to Phœnician traders. Egypt, during a period of 7,000 years, has preserved the same level, or at the most has sunk as slowly as the Nile mud has accumulated. Parts of the English and Scotch coast have risen perhaps twenty feet since the prehistoric period, when canoes were wrecked under what are now the streets of Glasgow, and whales were stranded in the Carse of Stirling. There is even some evidence that the latest rise may have occurred since the Roman wall was built from the Forth to the Clyde. In any case, however, the movements have been extremely slow, and there have been frequent oscillations, and long pauses when the level of land and sea remained stationary. The evidence, therefore, from the great changes which have occurred during each geological period, points to the same conclusion as that drawn from the thickness of formations, such as the coal measures and chalk, which must have been accumulated very slowly, viz., that geological time must be measured by a scale of millions of years.

Another test of the vast duration of geological time is afforded by the changes which have taken place in animal life as we pass from one formation to another, and even within the limits of the same formation. The fauna, or form of existing life at a given period, changes with extreme slowness. During the historical period there has been no perceptible change, and even since the Pliocene period, which cannot be placed at a less distance from us than 200,000 years, and probably at much more, the change has been very small. In the limited class of large land animals it has been considerable; but if we take the far more numerous forms of shell-fish and other marine life, the old species which have become extinct and the new ones which have appeared, do not exceed five per cent. of the whole. This is the more remarkable as great vicissitudes of climate and variations of sea-level have occurred during the interval. The whole of the Glacial period has come and gone, and Britain has been by turns an archipelago of frozen islands and part of a continent extending over what is now the German Ocean, and pushing out into the Atlantic up to the one hundred fathom line.

Reasoning from these facts, assuming the rate of change in the forms of life to have been the same formerly, and summing up the many complete changes of fauna which have occurred during the separate geological formations, Lyell has arrived at the conclusion that geology requires a period of not less than 200 millions of years to account for the phenomena which it discloses.

Long as the record is of geological time, it is only that of one short chapter in the volume of the history of the universe. Geology only begins when the earth had cooled down into a state resembling the present; when winds blew, rains fell, rivers and seas eroded rocks and formed deposits, and when the conditions were such that life became possible by the remains of which those deposits can be identified.

But before this period began, which may be called that of the maturity or middle age of our planet, a much vaster time must be allowed for the contraction and cooling of the vaporous ether or cosmic matter of which it is formed, into the state in which the phenomena of geology became possible. And if vast in the case of the earth, how much vaster must be the life periods of the larger planets, such as Jupiter, which from their much greater size cool and contract much more slowly, and are not yet advanced beyond the stage of intense youthful heat and glowing luminosity which was left behind by our earth a great many tens of millions of years ago! And how vastly vaster must be that of the sun, whose mass and volume exceed those of Jupiter in a far higher ratio than Jupiter surpasses the earth!

And beyond all this in a third degree of vastness come the life periods of those stars or distant suns, which we know to be in some cases as much as three hundred times larger than our sun, and not nearly so far advanced as it in the process of emergence from the fiery nebulous into the solar stage.

To give some idea of the vast intervals of time required for these changes, a few facts and figures may be given.

One of the latest speculations of mathematical science is that the rotation of the earth is becoming slower, or in other words the day becoming longer, owing to the retarding action of the tides, which act as a brake on a revolving wheel. If so, mathematical calculation shows that the effect of the reaction on the moon of this action of the moon on the earth, must be that as the earth rotates more slowly, the moon recedes to a greater distance. And *vice versâ*, when the earth rotated more rapidly the moon was nearer to it, until at length, when the process is carried back far enough, we arrive at a time when the moon was at the earth's surface and the length of the day about three hours. In this state of things the moon is supposed to have been thrown off from the earth, either by one great convulsion, or, more probably, by small masses at a time forming a ring like that of Saturn, which ended by coalescing into a single satellite. With the moon, which is the principal cause of the tides, so much nearer the earth, their rise and fall must have been something enormous, and huge tidal waves like the bore of the Bay of Fundy, but perhaps 500 or 1,000 feet high, must have swept twice during each revolution of the earth on its axis, *i.e.*, twice every three or four hours, along all the narrower seas and channels and over all except the mountainous lands adjoining.

Now these conclusions may be true or not as regards phases of the earth's life prior to the Silurian period, from which downwards geology shows unmistakably that nothing of the sort, or in the least degree approaching to it, has occurred. But what I wish to point out is that all this superstructure of theory rests on a basis which really does admit of definite demonstration and calculation.

Halley found that when eclipses of the sun, recorded in ancient annals, are compared with recent observations, a discrepancy is discovered in the rate of the moon's motion, which must have been slightly slower then than it is now. Laplace apparently solved the difficulty by showing that this was an inevitable result of the law of gravity, when the varying eccentricity of the earth's orbit was properly taken into account; and the calculated amount of the variation from this cause was shown to be exactly what was required to reconcile the observations. But our great English mathematician, Adams, having recently gone over Laplace's calculations anew, discovered that some factors in the problem had been omitted, which reduced Laplace's acceleration of the moon's motion by about one-half, leaving the other half to be explained by a real increase in the length of the sidereal day, or time of one complete revolution of the earth about its axis. The retardation required is one sufficient to account for the total accumulated loss of an hour and a quarter in 2,000 years ; or in other words, the length of the day is now more by about $\frac{1}{37}$th part of a second than it was 2,000 years ago.

At this rate it would require 168,000 years to make a difference of 1 second in the length of the day; 10,080,000 years for a difference of 1 minute ; and 604,800,000 years for a difference of 1 hour. The rate would not be uniform for the past, for as the moon got nearer it would cause higher tides and more retardation ; still, the abyss of time seems almost inconceivable to get back to the state in which the earth could have rotated in three hours and thrown off the moon.

It is right, however, to state that all mathematical calculations of time, based on the assumed rate at which cosmic matter cools into suns and planets, and these into solid and habitable globes, are in the highest degree uncertain. If the original data are right, mathematical calculation inevitably gives right conclusions. But if the data are wrong, or what is the same thing, partial and imperfect, the conclusions will, with equal certainty, be wrong also. Now in this case we certainly do not know "the truth, the whole truth, and nothing but the truth" respecting these processes. Take what is perhaps the most difficult problem presented by science—how the sun keeps up so uniformly the enormous amount of heat which it is constantly radiating into space. This radiation is going on in every direction, and the solar heat received by the earth is only that minute portion of it which is intercepted by our little speck of a planet. All the planets together receive less than one 230,000,000th part of the total heat radiated away by the sun and apparently lost in space. Knowing the amount of heat from the sun's rays received at the earth's surface in a given time, we can calculate the total amount of heat radiated from the sun in that time. It amounts to this, that the sun in each second of time parts with as much heat as would be given out by the burning of 16,436 millions of millions of tons of the best anthracite coal. And radiation certainly at this rate, if not a higher one, has been going on ever since the commencement of the geological record, which must certainly be reckoned by a great many tens of millions of years.

What an illustration does this afford of that apparent "waste of Nature" which made Tennyson "falter where he firmly trod" when he came to consider "her secret meaning in her deeds"!

Yet there can be no doubt that vast as these figures are, they are all the result of natural laws, just as we find the law of gravity prevailing throughout space at distances expressed by figures equally vast. The question is, what laws? The only one we know of at present at all adequate to account for such a generation of heat, is the transformation into heat of the enormous amount of mechanical force or energy, resulting from the condensation of the mass of nebulous matter from which the sun was formed, into a mass of its present dimensions. This is no doubt a true cause as far as it goes. It is true that as the mass contracts, heat would be, so to speak, squeezed out of it, very much as water is squeezed out of a wet sponge by compressing it. But it is a question whether it is the sole and sufficient cause. Mathematicians have calculated that even if we suppose the original cosmic matter to have had an infinite extension, its condensation into the present sun would only have been sufficient to keep up the actual supply of solar heat for about 15 millions of years. Of this a large portion must have been exhausted before the earth was formed as a separate planet, and had cooled down into a habitable globe. But even if we took the whole it would be altogether insufficient. All competent geologists are agreed in requiring at least 100 millions of years to account for the changes which have taken place in the earth's surface since the first dawn of life recorded in the older rocks.

Various attempts have been made to reconcile the discrepancy. For instance, it has been said that the constantly repeated impact of masses of meteoric and cometic matter falling into the sun must have caused the destruction of a vast amount of mechanical energy which would be converted into heat. This is true as far as it goes, but it is impossible to conceive of the sun as a target kept at a perpetual and uniform white heat for millions of years by a rain of meteoric bullets constantly fired upon it. More plausibly it is said that we know nothing of the interior constitution of the sun, and that its solid nucleus may be vastly more compressed than is inferred from the dimensions of its visible disc, which is composed of glowing flames and vapours. This also may be a true cause, but, after making every allowance, we must fall back on the statement that the continuance for such enormous periods of such an enormous waste of energy as is given out by the sun, though certainly explainable by laws of Nature, depends on laws not yet thoroughly understood and explained.

Even in the case, comparatively small and near to us, of the earth, the condition of the interior and the rate of secular cooling afford problems which as yet wait for solution. The result of a number of careful experiments in mines and deep sinkings shows that the temperature, as we descend below the shallow superficial crust which is affected by the seasons, *i.e.*, by the solar radiation, increases at the average rate of 1° Fahrenheit for every 60 feet of depth. That is the average rate, though it varies a good deal in different localities. Now, at this rate we should soon reach a depth at which all known substances would be melted.

But astronomical considerations, derived from the Precession of the Equinoxes, favour the idea that the earth is a solid and not a fluid body, and require us in any case to assume a rigid crust of not less than ninety miles in thickness. And if the whole earth below a thin superficial crust were in an ordinary state of fluidity from heat, it is difficult to see how it could do otherwise than boil, that is, establish circulating currents throughout its mass with disengagement of vapour, in which case the surface crust must be very soon broken up and melted down, just as the superficial crust of a red-hot stream of lava is, if an infusion of fresh lava raises the stream below to white heat, or as a thin film of ice would be if boiling water were poured in below it.

All we can say is, that the laws under which matter behaves under conditions of heat, pressure, chemical action, and electricity so totally different as must prevail in the interior of the earth, and *à fortiori* in that of the sun, are as yet very partially known to us.

In the meantime the safest course is to hold by those conclusions of geology which, as far as they go, depend on laws really known to us. For instance, the quantity of mud carried down in a year by the Ganges or Mississippi, is a quantity which can be calculated within certain approximate limits. We can tell with certainty how much the deposit of this amount of mud would raise an area, say of 100 square miles, and how long it would take, at this rate, to lower the area of India drained by the Ganges, a sufficient number of feet to give matter enough to fill up the Gulf of Bengal. And if among the older formations we find one, like the Wealden for instance, similar in character to that now forming by the Ganges, we can approximate from its thickness to the time that may have been required to form it.

In calculations of this sort there is no *theory*, they are based on positive facts, limited only by a certain possible amount of error either way. In short, the conclusions of geology, at any rate up to the Silurian period when the present order of things was fairly inaugurated, are approximate *facts* and not *theories*, while the astronomical conclusions are *theories* based on *data* so uncertain, that while in some cases they give results incredibly short, like that of 15 millions of years for the whole past process of the formation of the solar system, in others they give results almost incredibly long, as in that which supposes the moon to have been thrown off when the earth was rotating in three hours, while the utmost actual retardation claimed from observation would require 600 millions of years to make it rotate in twenty-three hours instead of twenty-four.

To one who looks at these discussions between

geologists and astronomers not from the point of view of a specialist in either science, but from that of a dispassionate spectator, the safest course, in the present state of our knowledge, seems to be to assume that geology really proves the duration of the present order of things to have been somewhere over 100 millions of years, and that astronomy gives an enormous though unknown time beyond in the past, and to come in the future, for the birth, growth, maturity, decline, and death of the solar system of which our earth is a small planet now passing through the habitable phase.

So far, however, as the immediate object of this work is concerned, viz., the bearings of modern scientific discovery on modern thought, it is not very material whether the shortest or longest possible standards of time are adopted. The conclusions as to man's position in the universe and the historical truth or falsehood of old beliefs, are the same whether man has existed in a state of constant though slow progression for the last 50,000 years of a period of 15 millions, or for the last 500,000 years of a period of 150 millions. It is a matter of the deepest scientific interest to arrive at the truth, both as to the age of the solar system, the age of the earth as a body capable of supporting life, the successive orders and dates at which life actually appeared, and the manner and date of the appearance of the most highly organised form of life endowed with new capacities for developing reason and conscience in the form of Man. Those who wish to prove themselves worthy of their great good luck in having been born in a civilised country of the nineteenth century, and not in Palæolithic periods, will do well to show that curiosity, or appetite for knowledge, which mainly distinguishes the clever from the stupid and the civilised from the savage man, by studying the works of such writers as Lyell, Huxley, Tyndall, and Proctor, where they will find the questions here only briefly stated, developed at fuller length with the most accurate science and in the clearest and most attractive style. But for the moral, philosophical, and religious bearings of these discoveries on the current of modern thought, there is such a wide margin that it becomes almost immaterial whether the shortest possible or longest possible periods should be ultimately established.

CHAPTER III.

MATTER.

Ether and Light—Colour and Heat—Matter and its Elements—Molecules and Atoms—Spectroscope—Uniformity of Matter throughout the Universe—Force and Motion—Conservation of Energy—Electricity, Magnetism, and Chemical Action—Dissipation of Heat—Birth and Death of Worlds.

WHAT is the material universe composed of? Ether, Matter, and Energy. Ether is not actually known to us by any test of which the senses can take cognizance, but is a sort of mathematical substance which we are compelled to assume in order to account for the phenomena of light and heat. Light, as we have seen, radiates in all directions from a luminous centre, travelling at the rate of 184,000 miles per second. Now what is light? It is a sensation produced on the brain by something which has been concentrated by the lens of the eye on the retina, and thence transmitted along the optic nerve to the brain, where it sets certain molecules vibrating. What is the *something* which produces this effect? Is it a succession of minute particles, shot like rifle-bullets from the luminous body and impinging on the retina as on a target? Or is it a succession of tiny waves breaking on the retina as the waves of the sea break on a shore?

Analogy suggests the latter, for in the case of the sister sense, Sound, we know as a fact that the sensation is produced on the brain by waves of air concentrated by the ear, and striking on the auditory nerve. But we have a more conclusive proof. If one of a series of particles shot out like bullets overtakes another, the force of impact of the two is increased; but if one wave overtakes another when the crest of the pursuing wave just coincides with the hollow of the wave before it the effect is neutralised, and if the two are of equal size it will be exactly neutralised and both waves will be effaced. In other words, two lights will make darkness. This, therefore, affords an infallible test. If two lights can make darkness, light is propagated, like sound, by waves. Now two lights do constantly make darkness, as is proved every day by numerous experiments. Therefore light is caused by waves.

But to have waves there must be a medium through which the waves are propagated. Without water you could not have ocean waves; without air you could not have sound-waves. Waves are in fact nothing but the successive forms assumed by a set of particles which, when forced from a position of rest, tend to return to that position, and oscillate about it. Place a cork on the surface of a still pond, and then throw in a stone; what follows? Waves are propagated, which seem to travel outwards in circles, but if you watch the cork, you will see that it does not really travel outwards, but simply rises and falls in the same place. This is equally true of waves of sound and waves of light. But the velocity with which the waves travel depends on the nature of the medium.

In a dense medium of imperfect elasticity they travel slowly, in a rare and elastic medium quickly. Now the velocity of a sound-wave in air is about 1,100 feet a second, that of the light-wave about 184,000 miles a second, or about one million times greater. It is proved by mathematical calculation that, if the density of two media are the same, their elasticities are in proportion to the squares of the velocities with which a wave travels. The elasticity of ether, therefore, would be a million million times greater than that of air, which, as we know, is measured by its power of resisting a pressure of about 15 lbs. to the square inch. But the ether must in fact be almost infinitely rare, as well as almost infinitely elastic, for it causes no perceptible retardation in the motions of the earth and planets. It must be almost infinitely rare also, because it permeates freely the interior of substances like glass and crystals, through which light-waves pass, showing that the atoms or ultimate particles of which these substances are composed, minute as they are, must be floating in ether like buoys floating on water or balloons in the air.

The dimensions of the light-waves which travel through this ether at the rate of 184,000 miles a second, can be accurately measured by strict mathematical calculations, depending mainly on the phenomena of interferences, *i.e.*, of the intervals required between successive waves for the crest of one to overtake the depression of another and thus make two lights produce darkness.

These calculations are much too intricate to admit of popular explanation, but they are as certain as those of the Nautical Almanac, based on the law of gravity, which enable ships to find their way across the pathless ocean, and they give the following results:

DIMENSIONS OF LIGHT-WAVES.

Colours.	Number of Waves in One Inch.	Number of Oscillations in One Second.
Red	39,000	477,000,000,000,000
Orange	42,000	506,000,000,000,000
Yellow	44,000	535,000,000,000,000
Green	47,000	575,000,000,000,000
Blue	51,000	622,000,000,000,000
Indigo	54,000	658,000,000,000,000
Violet	57,000	699,000,000,000,000

These are the colours whose vibrations affect the brain through the eye with the sensation of light, and which cause the sensation of white light when their different vibrations reach the eye simultaneously. But there are waves and vibrations on each side of these limits, which produce different effects, the longer waves with slower oscillations beyond the red, though no longer causing light causing heat, while the shorter and quicker waves beyond the violet cause chemical action, and are the most active agents in photography.

We must refer our readers to works treating specially of light for further details, and for an account of the vast variety of beautiful and interesting experiments with polarised light, coloured rings, and otherwise, to which the theory of waves propagated through ether affords the key. For the present purpose it is sufficient to say that modern science compels us to assume, as the substratum of the material universe, such an ether extending everywhere, from the faintest star seen at a distance which requires thousands of years for its rays, travelling at the rate of 184,000 miles a second, to reach the earth, down to the infinitesimally small interspace between the atoms of the minutest matter. And throughout the whole of this enormous range law prevails, ether vibrates and has always vibrated in the same definite manner, just as air vibrates by definite laws when the strings of a piano are struck by the hammers.

I pass now to the consideration of matter.

What is matter? In the most general sense it is that which has weight, or is subject to the law of gravity. The next analysis shows that it is something which can exist in the three forms of solid, liquid, or gas, according to the amount of heat. Diminish heat, and the particles approach closer and are linked together by mutual attraction, so as not to be readily parted; this is a solid. Increase the heat up to a certain point, and the particles recede until their mutual attractions in the interior of the mass neutralise one another, so that the particles can move freely, though still held together as a mass by the sum of all these attractions acting as if concentrated at the centre of gravity; this is the liquid state. Increase the heat still more, and the particles separate until they get beyond the sphere of their mutual attraction and tend to dart off into space, unless confined by some surface on which they exert pressure; this is a gas.

The most familiar instance of this is afforded by water, which, as we all know, exists in the three forms of ice, water, and vapour or steam, according to the dose of heat which has been incorporated with it.

Pursuing our inquiry further, the next great fact in regard to matter is that it is not all uniform. While most of the common forms with which we are conversant are made up of mixed materials, which can be taken to pieces and shown separately, there are, as at present ascertained, some seventy-one substances which defy chemical analysis to decompose them, and must therefore be taken as elementary substances. A great majority of these consist of substances existing in minute quantities, and hardly known outside the laboratories of chemists.

The world of matter, as known to the senses, is mainly composed of combinations, more or less complex, of a few elements. Thus, water is a compound of two simple gases, oxygen and hydrogen; air, of oxygen and nitrogen; the solid framework of the earth, mainly of combinations of oxygen with carbon, calcium, aluminum, silicon, and a few other bases; salt, of chlorine and sodium; the vegetable world directly and the animal world indirectly, mainly of complex combinations of oxygen, hydrogen, and nitrogen with carbon, and with smaller quantities of silicon, sulphur, potassium, sodium, and phosphorus. The ordinary metals, such as iron, gold, silver, copper, tin, lead, mercury, zinc, nearly complete the list of what may be called ordinary elements.

Now let us push our analysis a step further. How is matter made up of these elements? Up to and beyond the furthest point visible by aid of the microscope, matter is divisible. We can break a crystal into fragments, or divide a drop into drops, until they cease to be visible, though still retaining all the properties of the original substance. Can we carry on this process indefinitely, and is matter composed of something that can be divided and subdivided into fractional parts *ad infinitum?* The answer is, No, it consists of ultimate but still definite particles which cannot be further subdivided. How is this known? Because we find by experience that substances will only combine in certain definite proportions either of weight or measure. For instance, in forming water exactly eight grains by weight of oxygen combine with exactly one grain of hydrogen, and if there is any excess or fractional part of either gas, it remains over in its original form uncombined. In like manner, matter in the form of gas always combines with other matter in the same form by volumes which bear a definite and very simple proportion to each other, and the compound formed bears a definite and very simple ratio to the sum of the volumes of the combining gases. Thus two volumes of hydrogen combine with one of oxygen to form two volumes of water in the state of vapour.

From these facts certain inferences can be drawn. In the first place it is clear that matter really does consist of minute particles, which do not touch and form a continuous solid but are separated by intervals which increase with increase of temperature. This is evident from the fact that we can pour a second or third gas into a space already occupied by a first one. Each gas occupies the enclosed space just as if there were no other gas present, and exerts its own proper pressure on the containing vessel, so that the total pressure on it is exactly the sum of the partial pressures. It is easy to see what this means. If a second regiment can be marched into a limited space of ground on which a first regiment is already drawn up, it is evident that the first regiment must be drawn up in loose order, *i.e.*, the soldier-units of which it is composed must stand so far apart that other soldier-units can find room between them without disturbing the formation. But the effect will be that the fire from the front will be increased, as for instance if a soldier of the second regiment, armed with a six-shooter repeating rifle, takes his stand between two soldiers of the first regiment armed with single-barrelled rifles, the effective fire will be increased in the ratio of 8 to 2. And this is precisely what is meant by the statement that the pressure of two gases in the same space is the sum of the separate pressures of each. It is clearly established that the pressure of a gas on a containing surface is caused by the bombarding to which it is subjected from the impacts of an almost infinite number of these almost infinitely small atoms, which, when let loose from the mutual attractions which hold them together in the solid and fluid state, dart about in all directions, colliding with one another and rebounding, like a set of little billiard-balls gone mad, and producing a certain average resultant of momentum outwards which is called pressure.

Another simile may help us to conceive how the indivisibility of atoms is inferred from the fact that they only combine in definite proportions. Suppose a number of gentlemen and ladies promenading promiscuously in a room. The band strikes up a waltz, and they at once proceed to group themselves in couples rotating with rhythmical motion in definite orbits. Clearly, if there are more ladies than gentlemen, some of them will be left without partners. So, if instead of a waltz it were a threesome reel, in which each gentleman led out two ladies, there must be exactly twice as many ladies as gentlemen for all to join in the dance. But if a gentleman could be cut up into fractional parts, and each fraction developed into a dancing gentleman, as primitive cells split up and produce fresh cells, it would not matter how many ladies there were, as each could be provided with a partner. Now this is strictly analogous to what occurs in chemical combination. Water is formed by each gentleman atom of oxygen taking out a lady atom of hydrogen in each hand, and the sets thus formed commence to dance threesome reels in definite time and measure, any surplus oxygen or hydrogen atoms being left out in the cold. Wonderful as it may appear, science enables us not only to say of these inconceivably minute atoms that they have a real existence, but to count and weigh them. This fact has been accomplished by mathematical calculations based on laws which have been ascertained by a long series of experiments on the constitution of gases.

It is found that all substances, when in the form of gas, conform to three laws:

1. Their volume is inversely proportional to the pressure to which they are subjected.
2. Their volume is directly proportional to the temperature.
3. At the same pressure and temperature all gases have the same number of molecules in the same volume.

From the last law it is obvious that if equal volumes of two gases are of different weight, the cause must be that the molecules of the one are heavier than those of the other. This enables us to express the weight of the molecule of any other gas in some multiple of the unit afforded by the weight of the molecule of the lightest gas, which is hydrogen. Thus, the density of watery vapour being nine times that of hydrogen, we infer that the molecule of water weighs nine times as much as the molecule of hydrogen, and that of oxygen being eight times greater, we infer that the oxygen molecule is eight times heavier than that of hydrogen.

These weights are checked by the other law which has been stated, that chemical combination between different substances always takes place in certain definite proportions. Thus, whenever in a chemical process the original substances or the product are or might exist in the state of gas, it is always found that the definite proportions observed in the chemical process are either the proportions of the densities of the respective gases or some simple multiple of these proportions. Thus, the weight of hydrogen being 2, which combines with a weight of oxygen equal to 16 to form a weight of watery vapour equal to 18, the density of the latter is to that of hydrogen as 9 to 1, *i.e.*, as 18 to 2.

But to get to the bottom of the matter we must go a step further, and as we have decomposed substances into molecules, we must take the molecules themselves to pieces and see what they are made of. The molecule is the ultimate particle into which any substance can be divided retaining its own peculiar qualities. A molecule of water is as truly water as a drop or a tumblerful. But when chemical decomposition takes place, instead of the molecule of water we have molecules of two entirely different substances, oxygen and hydrogen. Nothing can well be more unlike than the product water and the component parts of which it is made up. Water is a fluid, oxygen a gas; water extinguishes fire, oxygen creates it. Water is a harmless drink, oxygen the base of the most corrosive acids. It is evident that the water-molecule is a composite, and that its qualities depend, not on the essential qualities of the atoms which have combined to make it, but on the manner of the combination, and the new modes of action into which these atoms have been forced. In his native war-paint oxygen is a furious savage; with a hydrogen atom in each hand he is a polished gentleman.

Our theory, therefore, leads beyond molecules to atoms, and we have to consider these particles of a still smaller order than molecules, as the ultimate indivisible units of matter of which we have been in search. And even these we must conceive of as corks, as it were, floating in an ocean of ether, causing waves in it by their own proper movements, and agitated by all the successive waves which vibrate through this ether-ocean in the form of light and heat.

Working on these data, a variety of refined mathematical calculations made by Clausius, Clark Maxwell, Sir W. Thomson, and other eminent mathematicians, have given us approximate figures for the actual size, weight, and velocities of atoms and molecules. The results are truly marvellous. A millimetre is the one-thousandth part of a metre, or roughly one twenty-fifth of an inch. The magnitudes with which we have to deal are all of an order where the standard of measurement is expressed by the millionth part of a millimetre. The volume of a molecule of air is only a small fraction of that of a cube whose side would be the millionth of a millimetre. A cubic centimetre, or say a cube whose side is between one-third and one-half of an inch, contains 21,000,000,000,000,000,000,000 molecules. The number of impacts received by each molecule of air during one second will be 4,700 millions. The distance traversed between each impact averages 95 millionths of a millimetre.

It may assist in forming some conception of these almost infinitely small magnitudes, to quote an illustration given by Sir W. Thomson as the result of mathematical calculation. Suppose a drop of water were magnified so as to appear of the size of the earth or with a diameter of 8,000 miles, the atoms of which it is composed, magnified on the same scale, would appear of a size intermediate between that of a rifle-bullet and of a cricket-ball.

These figures show that space and magnitude extend beyond the standards of ordinary human sense, such as miles, feet, and inches, as far downwards into the region of the infinitely small as they do upwards into that of the infinitely great.

And throughout the whole of this enormous range law prevails. The same law of gravity gives weight to molecules and atoms, makes an apple fall to the ground, and causes double stars to revolve round their centre of gravity in elliptic orbits. The law of polarity which converts iron-filings into small magnets under the influence of a permanent magnet or electric current, animates the smallest atom. Atoms arrange themselves into molecules, and molecules into crystals, very much as magnetised iron-filings arrange themselves into regular curves. And the great law seems to prevail universally throughout the material, as it does also throughout the moral world, that you cannot have a North without a South Pole, a positive without a negative, a right without a wrong; and that error consists mainly in what the poet calls "the falsehood of extremes"—that is, in allowing the attraction of one pole, or of one opinion, so to absorb us as to take no account of its opposite.

The universal prevalence of law has received wonderful confirmation of late years from the discovery made by the spectroscope that the sun, the planets, and the remotest stars are all composed of matter identical with that into which chemical analysis has resolved the constituent matter of the earth. This has been proved in the following way:

If a beam of light is admitted into a darkened room through a small hole or narrow slit, and a triangular piece of glass, called a prism, is interposed in its path, the image thrown on a screen is a rainbow-tinted streak, intersected by numerous fine dark lines, which is called a spectrum. If, instead of solar light, light from other luminous sources is similarly treated, it is found that all elementary substances have their peculiar spectra. Light from solid or liquid substances gives a continuous spectrum, light from gases or glowing vapours gives a spectrum of bright lines separated from each other, but always in definite positions according to the nature of the substance. The next great step in the discovery was that these bright lines become dark lines when a light of greater intensity, coming from a solid nucleus, is transmitted through an atmosphere of such gases or vapours. We can thus photograph the spectrum of glowing hydrogen, sodium, iron, or other substances, and placing it below a photograph of a solar or stellar spectrum, see if any of the dark lines of the latter correspond with the bright lines of the former. If they do we may be certain that these substances actually exist in the sun or star. It is, in fact, just the same thing as if we had been able to bring down a jar full of the solar or stellar matter and analyse it in our laboratories.

It is difficult to convey any adequate description of these grand discoveries made by the new science of Spectroscopy without referring to special works on the subject; but it may be possible to give some general idea of the principles on which they are based.

Light consists of waves propagated through ether. These waves are started by the vibrations of the ultimate particles of matter, which, whether in the simplest form of atoms, in the more complex form of molecules, or in the still more complex form of compound molecules, have their own peculiar and distinct vibrations. These vibrations are increased, diminished, or otherwise modified by variations of heat and by the collisions which occur between the particles from their own proper motions. If we take the simplest case, that of matter in the form of a gas or vapour composed of single atoms, at a temperature just sufficient to become luminous and at a pressure small enough to keep the atoms widely apart, the vibrations are all of one sort, viz., that peculiar to the elementary substance to which they belong, and one set of waves only is propagated by them through the ether. The spectrum, therefore, of such a gas is a single line of light, in the definite position which is due to its refrangibility, *i.e.*, to the velocity of the particular wave of light which the particular vibration of those particular atoms is able to propagate.

When pressure is increased so that the particles are brought closer together, their vibrations made more energetic and their collisions more frequent, more waves, and waves of different qualities are started, and more lines appear in the spectrum and the lines widen out, until at length when the gas becomes very dense, some of the lines overlap and an approach is made towards a continuous spectrum. Finally, when the particles are brought so near together that the substance assumes a fluid or solid state, the number of wave-producing vibrations becomes so great that a complete system of different light-waves is propagated, and the lines of the spectrum are multiplied until they coalesce and form a continuous band of rainbow-tinted light. If the particles of the gas, instead of being single atoms, are more complex, as molecules or compound molecules, the vibrations are more complex and the different resulting light-waves more numerous, so that the lines in the spectrum are more numerous, and in some cases they coalesce so as to form shaded bands, or what are called fluted lines, instead of simple lines.

Moreover, whatever light-waves are originated by the vibrations of the particles of a gas are absorbed into those vibrations and extinguished, if they originate from the vibrations of some more energetic particles of another substance outside of it, whose light-waves, travelling along the ether, pass through the gas, and are thus shown as dark lines in the spectrum of the other source of light.

We can now understand how the assertion is justified that we can analyse the composition of the sun and stars as certainly as if we had a jar full of their substance to analyse in our laboratory. The first glance at a spectrum tells us whether the luminous source is solid, fluid, or gaseous. If its spectrum is continuous it is solid or fluid; we know this for certain, but can tell nothing more. But if it consists of bright lines, we know that it comes direct from matter in the form of luminous gas, and knowing from experiments in the laboratory the exact colours and situations of the lines formed by the different elements of which earthly matter is composed, we can see whether the lines in the spectra of heavenly matter do or do not correspond with any of them. If bright lines correspond we are sure that the substances correspond, both as to their elementary atoms and their condition as glowing gas. If dark lines in the spectrum of the heavenly body correspond with bright lines in that of a known earthly substance, we are certain that the substances are the same and in the same state of gas, but that the solar or stellar spectrum proceeds from an intensely heated interior solid or fluid nucleus, whose waves have passed through an outer envelope or atmosphere of this gas.

Applying these principles, although the science is still in its infancy and many interesting discoveries remain to be made, this grand discovery has become an axiomatic fact—Matter is alike everywhere. The light of stars up to the extreme boundary of the visible universe, is composed mainly of glowing hydrogen, the same identical hydrogen as we get by decomposing water by a voltaic battery.

Of the 71 elementary substances of earthly matter enumerated by chemists, 9 may be considered as doubtful or existing only in excessively minute quantities. Of the remaining 62, 22 are known certainly to exist in the sun's atmosphere, 10 more can probably be traced there, and there are only 6 as to which, in the present state of our knowledge, there is negative evidence that they are not present. The elements whose presence is proved comprise many of those which are most common in the composition of the earth, as hydrogen, iron, lead, calcium, aluminium, magnesium, sodium, potassium, etc.; and if others, such as oxygen, carbon, and chlorine have not yet been found, good reasons may be assigned why they may not exist in a state likely to give recognisable spectrum-lines. The main fact is firmly established that matter is the same throughout all space, from the minutest atom to the remotest star.

Thus far we have been treating of matter only, and of force and motion but incidentally. These, however, are equally essential components of the phenomena of the universe. What is force? In the last analysis it is the unknown cause which we assume for motion, or the term in which we sum up whatever produces or tends to produce it. The idea of force, like so many other of our ideas, is taken from our own sensations. If we lift a weight or bend a bow, we are conscious of doing so by an effort. Something which we call will produces a motion in the molecules of the brain, which is transmitted by the nerves to the muscles, where it liberates a certain amount of energy stored up by the chemical composition and decomposition of the atoms of food which we consume. This contracts the muscle, and the force of its contraction, transmitted by a system of pulleys and levers to the hand, lifts the weight. If we let go the weight it falls, and the force which lifted it reappears in the force with which it strikes the ground. If we do not let go the weight but place it on a support at the height to which we have raised it, it does not fall, no motion ensues, but the lifting force remains stored up in a tendency to motion, and can be made to reappear as motion at any time by withdrawing the support, when the weight will fall. It is evident, therefore, that force may exist in two forms, either as actually causing motion or as causing a tendency to motion.

In this generalised form it has been agreed to call it energy, as less liable to be obscured by the ordinary impressions attached to the word force, which are mainly derived from experiences of actual motion cognizable by the senses. We speak, therefore, of energy as of something which is the basis or *primum mobile* of all motion or tendency to motion, whether it be in the grosser forms of gravity and mechanical work, or in the subtler forms of molecular and atomic motions causing the phenomena of heat, light, electricity, magnetism, and chemical action. This energy may exist either in the form of actual motion, when it is called energy of motion, or in that of tendency to motion, when it is called energy of position. Thus the bent bow has energy of position which, when the string is let go, is at once converted into energy of motion in the flight of the arrow.

Respecting this energy modern science has arrived at this grand generalisation, that it is one and the same in all its different manifestations, and can neither be created nor destroyed, so that all these varied manifestations are mere transformations of the same primitive energy from one form to another. This is what is meant by the principle of the "Conservation of Energy."

It was arrived at in this way. Speaking roughly

it has long been known that heat could generate mechanical power, as seen in the steam-engine; and conversely that mechanical power could generate heat, as is seen when a sailor, in a chill north-easter, claps his arms together on his breast to warm himself. But it was reserved for Dr. Joule to give this fact the scientific precision of a natural law, by actually measuring the amount of heat that was added to a given weight of water by a given expenditure of mechanical power, and conversely the amount of mechanical work that could be got from a given expenditure of heat.

A vast number of carefully-conducted experiments have led to the conclusion that if a kilogramme be allowed to fall through 424 metres and its motion be then suddenly stopped, sufficient heat will be generated to raise the temperature of one kilogramme of water by 1° Centigrade; and conversely this amount of heat would be sufficient to raise one kilogramme to a height of 424 metres.

If, therefore, we take as our unit of work that of raising one kilogramme one metre, and as our unit of heat that necessary to raise one kilogramme of water 1° Centigrade, we may express the proportion of heat to work by saying that one unit of heat is equal to 424 units of work; or, as it is sometimes expressed, that the number 424 is the mechanical equivalent of heat.

But the question may be asked, what does this mean, how can mechanical work be really transformed into heat or *vice versâ?* The answer is, the energy which was supplied by chemical action to the muscles of the man or horse, or to the water converted into steam by combustion of coal, which originated the mechanical work, was first transformed into its equivalent amount of mechanical energy of motion, and then, when that motion was arrested, was transformed into heat, which is simply the same energy transformed into increased molecular motion.

If we wish to carry our inquiry a step further back and ask where the original energy came from which has undergone these transformations, the answer must be, mainly from the sun. The sun's rays, acting on the chlorophyl or green matter of the plants of the coal era, tore asunder the atoms of carbon and oxygen which formed the carbonic acid in the atmosphere, and locked up a store of energy in the form of carbon in the coal which is burned to produce the steam. In like manner it stored up the energy in the form of carbon in the vegetable products which, either directly, or indirectly after having passed through the body of some animal, supplied the food, whose slow combustion in the man or horse supplied the energy which did the work.

But where did the energy come from which the sun has been pouring forth for countless ages in the form of light and heat, and of which our earth only intercepts the minutest portion? This is a mystery not yet completely solved, but one real cause we can see, which has certainly operated and perhaps been the only one, viz., the mechanical energy of the condensation by gravity of the atoms which originally formed the nebulous matter out of which the sun was made. If we ask how came the atoms into existence endowed with this marvellous energy, we have reached the furthest bounds of human knowledge, and can only reply in the words of the poet: "Behind the veil, behind the veil."

We can only form metaphysical conceptions, or I might rather call them the vaguest guesses. One is, that they were created and endowed with their elementary properties by an all-wise and all-powerful Creator. This is Theism.

Another, that thought is the only reality, and that all the phenomena of the universe are thoughts or ideas of one universal, all-pervading Mind. This is Pantheism.

Or again, we may frankly acknowledge that the real essence and origin of things are "behind the veil," and not knowable or even conceivable by any faculties with which the human mind is endowed in its present state of existence. This is Agnosticism.

There is one other conception, of which we may certainly say that it is not true—that is Atheism. No one with the least knowledge of science can maintain that it can ever be demonstrated that everything in the universe exists of itself and never had a Creator.

But these speculations lead us into the misty regions where, like Milton's devils, "we find no end in wandering mazes lost." Let us return to the solid ground of fact, on which alone the human mind can stand firmly, and like Antæus gather fresh vigour every time it touches it for further efforts to enlarge the boundaries of knowledge and extend the domain of Cosmos over Chaos.

The transformation of energy which we have seen to exist in the case of mechanical work and heat, is not confined to those two cases only, but is a universal law applicable to all actions and arrangements of matter which involve motions of atoms, molecules, or masses, and therefore imply the existence of energy. In heat we have had an example of energy exerted in molecular motion and molecular separation. In chemical action we have energy exerted in the separation of atoms, severing them from old combinations and mutual attractions, and bringing them within the sphere of new ones. In electricity, and magnetism which is another form of electricity, we have energy of position which manifests itself in electrical separation, by which matter becomes charged with two opposite energies, positive and negative, which accumulate at separate poles, or on separate surfaces, with an amount of tension which may be reconverted into the original amount of energy of motion when the spark, passing between them, restores their electrical equilibrium. Of this we have an example in the ordinary electrical machine, where the original energy comes from the mechanical force which turns the handle, and is given back when the electric spark brings things back to their original state.

We have also energy of motion, when instead of electrical separation and tension we have a flow or current of electricity producing the effect of the electric spark in a slow, quiet, and continuous manner. Thus, in the voltaic battery, the free energy created by the difference of chemical action of an acid on plates of different metals, is transformed into a current which charges two poles with opposite electricities, and when the poles are brought together and the circuit is closed, flows through it in a continuous current. This current is an energetic agent which produces various effects. It deflects the magnetic needle, as is seen in the electric telegraph. It creates magnetism, as is seen when the poles of the battery are connected by a wire wrapped round and round a cylinder of soft iron, so as to make the current circulate at right angles to the axis formed by the cylinder. In fact, all magnetism may be considered as the summing up at the two opposite extremities or poles of an axis, of the effects of electric currents circulating round it ; as, for instance, the earth is a great magnet because currents caused by the action of the sun circulate round it nearly parallel to the equator. Electric currents further show their energy by attracting and repelling one another, those flowing in the same direction attracting, and those in opposite directions repelling, the same effect showing itself in magnets, which are in substance collections of circular currents flowing from right to left or left to right according as they are positive or negative. Again, currents produce an effect by inducing currents in other bodies placed near them, very much as the vibrations of a tuning-fork induce vibrations and bring out a corresponding note from the strings of a piano or violin ready to sound it. When a coil of wire is connected with a battery and a current passes through it, if it is brought near to another isolated coil it induces a current in an opposite direction, which, when it recedes from it, is changed into a current in the same direction.

These principles are illustrated by the ordinary dynamo, by which the energy of mechanical work exerted in making magnets revolve in presence of currents, and by various devices accumulating electric energy, is made available either for doing other mechanical work, such as driving a wheel, or for doing molecular or atomic work by producing heat and light.

For another transformation of the energy of electric currents is into heat, light, or chemical action. If the two poles of a battery are connected by a thin platinum wire it will be heated to redness in a few seconds, the friction or resistance to the current in passing through the limited section of the thin wire producing great heat. If the wire is thicker heat will equally be produced, but more slowly.

If the poles of the battery are made of carbon, or some substance the particles of which remain solid during intense heat, when they are brought nearly together the current will be completed by an arc of intensely brilliant light, and the carbon will slowly burn away. This is the electric light so commonly used when great illuminating power is wanted.

Again, the electric current may employ its energy in effecting chemical action. If the poles of a battery, instead of being brought together, are plunged into a vessel of water, decomposition will begin. Oxygen will rise in small bubbles at the positive pole, and hydrogen at the negative. If these two gases are collected together in the same vessel, and an electric current, in the intense and momentary form of a spark, passed through them, they will combine with explosion into the exact amount of water which was decomposed in their formation.

Everywhere, therefore, we find the same law of universal application. Energy, like matter, cannot be created or destroyed, but only transformed. It is therefore, in one sense, eternal. But there is another point of view from which this has to be regarded.

Mechanical work, as we have seen, can always be converted into heat, and heat can, under certain conditions, be reconverted into mechanical work; but not under all conditions. The heat must pass from something at a higher temperature into something at a lower. If the condenser of a steam-engine were always at the same temperature as the boiler, we should get no work out of it. It is easy to understand how this is the case if we figure to ourselves a river running down into a lake. If the stream is dammed up at two different levels, each dam, as long as there is water in it, will turn a mill-wheel. But if all the water runs down into the lake and, owing to a dry season, there is no fresh supply, the wheels will stop and we can get no more work done. So with heat, if it all runs down to one uniform temperature it can no longer be made available to do work. In the case of the river, fresh water is supplied at the higher levels, by the sun's energy raising it by evaporation from the seas to the clouds, from which it is deposited as rain or snow. But in the case of heat there is no such self-restoring process, and the tendency is always towards its dissipation; or in other words, towards a more uniform distribution of heat throughout all existing matter. The process is very slow; the original fund of high-temperature heat is enormous, and as long as matter goes on condensing fresh supplies of heat are, so to speak, squeezed out of it.

Still there is a limit to condensation, while there is no limit to the tendency of heat to diffuse itself from hotter to colder matter until all temperatures are equalised. The energy is not destroyed; it is still there in the same average amount of total heat, though no longer differentiated into greater and lesser heats, and therefore no longer available for life, motion, or any other form of transformation. This seems to be the case with the moon, which, being so much smaller, has sooner equalised its heat with surrounding space, and is apparently a burnt-out and dried-up cinder without air or water. And this, as far as we see, must be the ultimate fate of all planets, suns, and solar systems. Fortunately the process is extremely slow, for even our small earth has enjoyed air, water, sunshine, and all the present conditions necessary for life for the whole geological period, certainly from the Silurian epoch downwards, if not earlier, which cannot well be less than 100 millions of years, and may be much more. Still time, even if reckoned by hundreds of millions of years, is not eternity; and as, looking through the telescope at nebulæ which appear to be condensing about central nuclei, we can dimly discern a beginning, so, looking at the moon and reasoning from established principles as to the dissipation of heat, we can dimly discern an end. What we really can see is that throughout the whole of this enormous range of space and time law prevails; that, given the original atoms and energies with their original qualities, everything else follows in a regular and inevitable succession; and that the whole material universe is a clock, so perfectly constructed from the beginning as to require no outside interference during the time it has to run to keep it going with absolute correctness.

CHAPTER IV.

LIFE.

Essence of Life—Simplest Form, Protoplasm—Monera and Protista — Animal and Vegetable Life — Spontaneous Generation — Development of Species from Primitive Cells—Supernatural Theory — Zoological Provinces — Separate Creations—Law or Miracle—Darwinian Theory—Struggle for Life—Survival of the Fittest — Development and Design — The Hand — Proof required to establish Darwin's Theory as a Law—Species—Hybrids—Man subject to Law.

THE universe is divided into two worlds — the inorganic, or world of dead matter; and the organic, or world of life. What is life? In its essence it is a state of matter in which the particles are in a continued state of flux, and the individual existence depends, not on the same particles remaining in the same definite shape, but on the permanence of a definite mould or form through which fresh particles are continually entering, forming new combinations and passing away. It may assist in forming a conception of this if we imagine ourselves to be looking at a mountain the top of which is enveloped in a driving mist. The mountain is dead matter, the particles of which continue fixed in the rocks. But the cloud form which envelops it is a mould into which fresh particles of vapour are continually entering and becoming visible on the windward side, and passing away and disappearing to leeward. If we add to this the conception that the particles do not, as in the case of the cloud, simply enter in and pass away without change, but are digested, that is, undergo chemical changes by which they are partly assimilated and worked up into component parts of the mould, and partly thrown off in new combinations, we shall arrive at something which is not far off the ultimate idea of what constitutes living matter, in its simplest form of the protoplasm, or speck of jelly-like substance, which is shown to be the primitive basis or raw material of all the more complex forms both of vegetable and animal life. Digestion, therefore, is the primary attribute. A crystal grows from *without*, by taking on fresh particles and building them up in regular layers according to fixed laws, just as the pyramids of Egypt were built up by laying layer upon layer of squared stones upon surfaces formed of regular figures, and inclined to each other at determinate angles.

The living plant or animal grows from within by taking supplies of fresh matter into its inner laboratory, where it is worked up into a variety of complex products needed for the existence and reproduction of life. After supplying these, the residue is given back in various forms to the inorganic world, and the final residue of all is given back by death, which is the ultimate end of all life.

The simplest form of life, in which it first emerges from the inorganic into the organic world, consists of protoplasm, or, as it has been called, the physical basis of life. Protoplasm is a colourless semi-fluid or jelly-like substance, which consists of albuminoid matter, or in other words, of a heterogeneous carbon-compound of very complex chemical composition. It exists in every living cell, and performs the functions of nutrition and reproduction, as well as of sensation and motion. In its simplest form, that of the microscopic monera or protista, the lowest of living beings, we find a homogeneous structureless piece of protoplasm, without any differentiation of parts. The monera are simple living globules of jelly, without even a nucleus or any sort of organ, and yet they perform all the essential functions of life without any different parts being told off for particular functions. Every particle or molecule is of the same chemical composition and a facsimile of the whole body, as in the case of a crystal. They are, therefore, the first step from the inorganic into the organic world, and if spontaneous generation takes place anywhere, it is in the passage of the chemical elements from the simple and stable combinations of the former into the complex and plastic combinations of the latter.

These monera are found principally in the sea and in great masses at the bottom of deep oceans, where they form a sort of living slime first described by Huxley in 1868, and called Bathybius.

The next step upwards is to the cell in which the protoplasm is enclosed in a skin or membrane of modified protoplasm, and a nucleus, or denser spot, is developed in the enclosed mass. This is the primary element from which all the more complicated forms of life are built up. Each cell seems to have an independent life of its own, and a faculty of reproduction by splitting into fresh cells similar to itself, which multiply in geometrical progression, assimilating the elements of their substance from the inorganic world so rapidly as to provide the requisite raw material for higher structures.

The first organised living forms are **extremely minute**, and can only be recognised by powerful microscopes. A filtered infusion of hay, allowed to stand for two days, will swarm with living things, a number of which do not exceed $\frac{1}{10,000}$th of an inch in diameter. Minute as these animalcula are, they are thoroughly alive. They dart about and digest; the smallest speck of jelly-like substance shoots out branches or processes to seize food, and if these come in collision with other substances they withdraw them. They exist in countless myriads, and perform a very important part in the economy of nature. They are the scavengers of the universe, and remove the remains of living matter after death, which would otherwise accumulate until they choked up the earth. This they do by the process of putrefaction, which is due mainly to the multiplication of little rod-like creatures known as bacteria, which work up the once living, now dead, matter into fresh elements, again fitted to play their part in the inorganic and organic worlds.

One of the simplest of these forms is the amœba, which is nothing but a naked little lump of cell-matter, or plasma, containing a nucleus; and yet this little speck of jelly moves freely, it shoots out tongues or processes and gradually draws itself up to them with a sort of wave-like motion; it eats and grows, and in growing reproduces itself by contracting in the middle and splitting up into two independent amœbæ.

The germs of these various animalcula swarm in the air, and carry seeds of infection everywhere where they find a soil fitted to receive them; and thus assist the survival of the fittest in the struggle of life, by eliminating weak and unhealthy individuals and species. Thus when the potato, the vine, or the silk-worm has had its constitution enfeebled by prolonged artificial culture, there are germs always ready to revenge the violation of natural laws, and bring the survivors back to a more healthy condition. In like manner the germs of cholera, typhoid, and scarlet fever, enforce the observance of sanitary principles.

In this simple form the lowest forms of life are not yet sufficiently differentiated to enable us to dis-

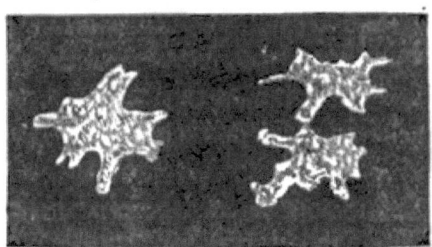

AMŒBA.   AMŒBA dividing into two.

tinguish clearly between animal and vegetable, and they have been called by some naturalists Protista, while others designate them as Protozoa or Protophyta, according as they show more resemblance to one or the other form of life. But it is often so doubtful that in looking at the same organism through a microscope, Huxley was inclined to consider it as a plant, while Tyndall exclaimed that he could as soon believe that a sheep was a vegetable.

In the next stage upwards, however, life subdivides itself into two great kingdoms, that of the vegetable and of the animal world. Alike in their general definition as contrasted with inorganic matter, and in their common origin from an embryo cell, which divides and subdivides until cell-aggregates are formed, from which the living form is built up by a process of evolution, the plant differs from the animal in this: that the former feeds directly on inorganic matter, while the latter can only feed on it indirectly, after it has been manufactured by the plant into vegetable substance.

This is universally true, for if we dine on beef, we dine practically on the grass which the ox ate; that is, on the carbon, oxygen, hydrogen, and other simple elements which the grass, under the stimulus of light and sunshine, manufactured into complex compounds; and which the ox again, by a second process, manufactured from these compounds into others still more complex, and more easily assimilated by us in the process of digestion. But in no case can we dine, as the plant does, on the simple elements, and thrive on a diet of air and water, with a small admixture of nitrate of ammonia, and of phosphates, sulphates and chlorides, of a few primitive metals. Vegetable life, therefore, is the producer, and animal life the consumer, of the organic world.

Practically the plant derives most of its substance from the carbonic acid gas in the atmosphere, which green leaves under the stimulus of light and heat have the faculty of decomposing, and abstract the carbon giving out the oxygen; while the animal, by a reverse process, burns up the compounds manufactured by the plant, principally out of this carbon, by the oxygen obtained from the air by the process of respiration, exhaling the surplus carbon in the form of carbonic acid gas.

The balancing effect of these two processes may be seen in any aquarium, where animals and vegetables live together in water which is kept pure, while it would become stagnant and poisonous in a few hours, if one of the two forms of life were removed. All that the animal requires therefore for its existence, materials with which to build up its frame and supply waste; heat with which to maintain its circulating fluids and other substances at a proper temperature; motive power or energy to enable it to move, feel, and in the case of man to think; are all proceeds of the slow combustion of materials derived from the vegetable world in the oxygen breathed from the air, just as the work done by a steam-engine is the product of a similar combustion, or chemical combination of the oxygen of the air with the coal shovelled into the fire-box. These distinctions, however, between animals and vegetables are not quite absolute, for, even in the more highly-organised forms of life, there is a border-land where some plants seem to perform the functions of animals, as in those which catch and consume flies and eat and digest pieces of raw meat.

Those who wish to pursue this interesting subject further will do well to read the Chapter on Living Matter in Huxley's "Physiography," where they will find it more fully explained, with the inimitable clearness which characterises all the writings of an author who is at the same time one of the first scientific authorities and one of the greatest masters of English prose. But my present object is not to write a scientific treatise, but shortly to sum up the ascertained results of modern science, with a view to their bearings on modern thought; and from this point of view the immediate question is, how far law, which has been shown to prevail universally throughout space, time, and inorganic matter, can be shown to prevail equally throughout the world of life.

Up to a certain point this admits of positive proof. It is as certain that all individual life, from the most elementary protoplasm up to the highest organism Man, originates in a minute or embryo cell, as it is that oxygen and hydrogen combined in certain proportions make water. But if we try to go back one step further, behind the cell, we are stopped. In the inorganic world we can reason our way beyond the microscopic matter to the molecule, and from the molecule to the atom, and are only arrested when we come to the ultimate form of matter, and of energy, out of which the universe is built up. But, in the case of life, we are stopped two steps short of this, and cannot tell how the cell containing the germ of life is built up out of the simpler elements.

Many attempts have been made to bridge over this gulf, and show how life may originate in chemical compounds, but hitherto without success. Experiments have been made which, for a time, seemed to show that spontaneous generation was a scientific fact, *i.e.*, that the lowest forms of life, such as bacteria and amœba, really did originate in infusions containing no germs of life; but they have been met by counter experiments confirming Harvey's *dictum*, "Omne animal ex ovo," or all life proceeds from antecedent germs of life, and the verdict of the best authorities, such as Pasteur, Tyndall, and Huxley is, that spontaneous generation has been "defeated along the whole line." This verdict is perhaps too unqualified, for it certainly appears that, on the assumption with which both sides started, that all organic life was destroyed by exposure to a heat of 212°, or the boiling-point of water, the advocates of spontaneous generation had the best of it, as low forms of life did appear in infusions which had been exposed to this heat, and then hermetically sealed, so as to prevent any germs from entering. But it was replied that, as a hard pea takes more boiling than a soft one, it might very well be that heat sufficient to destroy life in any moist organism of sufficient size to be seen by the microscope, might not destroy the germinating power of ultra-microscopic germs in a very dry state. And this position seems to have been confirmed by various experiments, showing that such ultra-microscopic germs really do exist, and are given forth in the last life stage of the bacteria which cause putrefaction; and that if they are absent or destroyed by repeated applications of heat, infusions will keep sweet for ever in optically pure air.

Above all, the germ theory has received confirmation from the brilliant practical results to which it has led in the hands of Pasteur, enabling him to detect, and to a great extent eradicate, the causes which had led to the oidium of the vine and the pebrine of the silk-worm, thereby saving losses of millions to the industries of France. The germ theory has also led to important results in medical science, and is pointing towards the possibility of combating the most fatal diseases by processes analogous to that by which vaccination has almost freed the human race from the scourge of small-pox.

On the whole, therefore, we must be content to accept a verdict of "Not proven" in the case of spontaneous generation, and admit that as regards the first origin of life, science fails us, and there is at present no known law that will account for it.

Should spontaneous generation ever be proved to be a fact, it will doubtless be in creating living protoplasm from inorganic elements at its earliest stage, before it has been differentiated even into the primitive form of a nucleated cell or that of an amœba. This is what the doctrine of evolution would lead us to expect, for it would be in contradiction to it to suppose that the starting-point could be interpolated at any stage subsequent to the lowest. It may be also that this step could only be made under conditions of heat, pressure, and otherwise, which existed in the earlier stage of the earth's existence, but have long since passed away.

This, however, is only a small part of the difficulty we have to encounter in reducing life to law.

These primeval embryo cells, like as they are in appearance, contain within them the germs of an almost infinite diversity of evolutions, each running its separate course distinct from the others. The world of life is not one and uniform, but consists of a vast variety of different species, from the speck of protoplasm up to the forest tree, and from the humble amœba up to man, each one, at any rate within long intervals of time, breeding true and keeping to its own separate and peculiar path along the line of evolution.

The first germ, or nucleated cell, of a bacteria develops into other bacteria and nothing else, that of a coral into corals, of an oak into oaks, of an elephant into elephants, of a man into man. In the latter case we can trace the embryo in its various stages of growth through forms having a certain analogy to those of the fish, the reptile, and the lower mammals, until it finally takes that of the human infant. But we have no experience of a fish, a frog, or a dog, being ever born of human parents, or of any of the lower animals ever producing anything resembling a man.

How can this be explained? Naturally the first attempt at explanation was by miracle. At a time when everything was explained by miracle, when all unusual occurrences were attributed to supernatural agency, and men lived in an atmosphere of providential interferences, witchcraft, magic, and all sorts of divine and diabolic agencies, nothing seemed easier than to say the beasts of the field, the birds of the air, and the fishes of the sea are all distinct after their kind, because God created them so.

But as the supernatural faded away and disappeared in other departments where it had so long reigned supreme, and science began to classify, arrange, and accumulate facts as they really are, it became more and more difficult, or rather impossible, to accept this simple explanation. The very first step destroyed the validity of all the traditional myths which described the origin of life from one simultaneous act of creation at a single centre. The earth is divided into separate zoological provinces, each with its own peculiar animal and vegetable world. The kangaroo, for instance, is found in Australia and there only. By no possibility could the aboriginal kangaroo have jumped at one bound from Mount Ararat to Australia, leaving no trace of his passage in any intermediate district. This isolation of life in separate provinces applies so rigidly, that we may sum it up by saying generally that there are no forms of life common to two provinces unless where migration is possible, or has been possible in past geological periods.

In islands at a distance from continents, we find common forms of marine life, for the sea affords a means of communication; and often common forms of bird, insect, and vegetable life, where they may have been wafted by the winds; but forms which neither in the adult or germ state could swim or fly, or be transported by something which did swim or fly, are invariably wanting. New Zealand affords a most conspicuous instance of this. Here is a large country with a soil and climate exceptionally well adapted to support a large amount of animal life of the higher orders, and yet it had absolutely no land animals before they were introduced by man. If special creations took place to replenish the earth as soon as any portion of its surface becomes fit to sustain it, why were there no animals in New Zealand? Or, in the Andaman Islands, in the Gulf of Bengal, which are as large as Ireland, covered with luxuriant vegetation, and within 300 miles of the coast of Asia, where similar jungles swarm with elephants, tigers, deer, and all the varied forms of mammalian life, there are no mammalia except a pigmy black savage and a pigmy black pig, the latter probably introduced by man.

The sharpness of the division between zoological provinces is well illustrated by that drawn by the Straits of Lombok, where a channel, not twenty miles wide, separates the fauna of Asia and Australia so completely that there are no species of land animals, and only a few of birds and insects, common to the two sides of a channel not so wide as the Straits of Dover.

There is no possibility of accounting for this, except

by supposing that the deep water fissure of the Strait of Lombok has existed from remote geological periods, and barred the migration southwards of those Asiatic animals, which, as long as they found dry land, migrated northwards and westwards till they were stopped by the Polar and Atlantic Oceans. This difficulty of requiring special creations for separate provinces is enormously enhanced if we look beyond the existing condition of things, and trace back the geological record. We must suppose separate creations for all the separate provinces of the separate successive formations from the Silurian upwards. And the more we investigate the conditions of life either under existing circumstances or in those of past geological epochs, the more enormously are we driven to multiply the number of separate creations which would be necessary to account for the diversity of species. We find life shading off into an infinite variety of almost imperceptible gradations from the highest organism, man, to the lowest, or speck of protoplasm, and we can draw no hard and fast line and say, up to this point life originated in law, and beyond it we must have recourse to miracle. Either all life or none is a product of evolution acting by defined law, and the affirmation of law is the negation of miracle.

Every day brings us an account of some new discovery bringing forms of life nearer together and bridging over intervals thought to be impassable. The discovery of plants living on insects, and which devour and digest pieces of raw meat, has added to the difficulty which has been long felt, in the humbler forms of life, of drawing any clear line of demarcation between the animal and vegetable worlds.

Microscopic research brings to light fresh facts confounding our fixed ideas as to the permanence of particular modes of reproducing life, and showing that the same organism may run through various metamorphoses in the course of its life-cycle, during some of which it may be sexual and in others asexual, *i.e.*, it may reproduce itself alternately by the co-operation of two beings of opposite sex, and by fissure or budding from one being only which is of no sex.

These, and a multitude of other similar facts, complicate enormously the problems of life and its developments, whether we attempt to solve it by calling in aid a perpetual series of innumerable miraculous interpositions, or by appealing to ordinary known laws of Nature.

Is the latter solution possible, and can the organic world be reduced, as the inorganic world has been with all its mysteries and infinities of space, time, and matter, from chaos into cosmos, and shown to depend on permanent and harmonious laws? Is the world of life, like that of matter, a clock, so perfectly constructed from the first that it goes without winding up or regulating? or is it a clock which would never have started going, or having started would soon cease to go, if the hand of the watchmaker were not constantly interfering with it? This is the question which the celebrated Darwinian theory attempts to answer, of which I now proceed to give a short general outline.

The varieties among domestic animals are obvious to every one. The race-horse is a very different creature from the dray-horse; the short-horned ox from the Guernsey cow; the greyhound from the Skye terrier. How has this come to pass? Evidently by man's intervention, causing long-continued selection in breeding for certain objects. The English race-horse is the product of mating animals distinguished for speed for some fifteen or twenty generations. The greyhound is a similar dog-product by breeding for a longer period with the same object; as the Skye terrier is of selection in order to get a dog which can follow a fox into a cairn of rocks and fight him when he gets there. In all these cases it is evident that the final result was not attained at once, but by taking advantage of small accidental variations and accumulating them from one generation to another by the principle of heredity, which makes offspring reproduce the qualities of their parents.

The most precise and scientific experiments on this power of integrating, or summing up, a progressive series of differentials, or minute differences, between successive generations, are those conducted by Darwin on pigeons. He has shown conclusively that all the races of domestic pigeons, of which there are two or three hundred, are derived from one common ancestor, the wild or blue rock pigeon, and that the pigeon-fancier can always obtain fresh varieties in a few generations by careful interbreeding. Of the existing varieties many now differ widely from one another, both in size, appearance, and even in anatomical structure, so that if they were now discovered for the first time in a fossil state or in a new country, they would assuredly be classed by naturalists as separate species.

This is the work of man; is there anything similar to it going on in Nature? Yes, says Darwin, there is a tendency in all life, and especially in the lower forms of life, to reproduce itself vastly quicker than the supply of food and the existence of other life can allow, and the balance of existence is only preserved by the wholesale waste of individuals in what may be called the "struggle for life." In this struggle, which goes on incessantly and on the largest scale, the slightest advantage must tell in the long run, and on the average, in selecting the few who are to survive, and such slight advantages must tend to accumulate from one generation to another under the law of heredity. The cumulative power of selection exercised by man in the breeding of races is therefore necessarily exercised in Nature by the struggle for life, and in the course of time, by the cumulation of advantages originally slight, small and fluctuating variations are hardened into large and permanent ones, and new species are formed.

Darwin illustrates this principle of the "struggle for life" with a vast variety of instances, showing how the balance of animal and vegetable life may be preserved or destroyed in the most unexpected manner. For instance, the fertilisation of red clover is effected by humble-bees, and depends on their number; the number of bees in a given district depends mainly on the number of field-mice which destroy their combs and nests; the number of mice depends on the number of cats; and thus the presence or absence of a carnivorous animal may decide the question whether a particular sort of flora shall prevail over others or be extirpated.

The countless profusion with which any one species, unchecked by its natural foes, may multiply in a given district, is illustrated by the potato disease, which in a few days invades whole countries; and by the rabbit plague in Australia and New Zealand, where, in less than twenty years, the descendants of a few imported pairs have rendered whole provinces useless for sheep pasture, and stoats are now being imported to restore the balance of life. The tendency in species to produce varieties which by selection may become exaggerated and fixed, is illustrated by the case of the Ancon herd of sheep. A ram lamb was born in Massachusetts in 1791, which had short crooked legs and a long back like a turnspit dog. Being unable to jump over fences like the ordinary sheep, it was thought to possess certain advantages to the farmer, and the breed was established by artificial selection in pairing this ram with its descendants who possessed the same peculiarities. The introduction of the Merino superseded the Ancon by giving a tame sheep not given to jump fences, with a better fleece, and so the breed was not continued, but it is certain that it might have been established as a permanent variety differing from the ordinary sheep as much as the turnspit or Skye terrier differs from the ordinary dog. The tendency of Nature to variation is apparent in the fact that of the many hundred millions of human beings living on the earth, no two are precisely alike, and varieties often appear, as in giants and dwarfs, six-fingered or toed children, hairy and other families, which might doubtless be fixed and perpetuated by artificial or natural selection, until they became strongly marked and permanent.

It is evident that if the theory of development is true it excludes the old theory of design, or rather, it thrusts it back in the organic, as it has been thrust back in the inorganic world, to the first atoms or origins which were made so perfect as to carry within them all subsequent phenomena by necessary evolution. Design and development lead to the same result, that of producing organs adapted for the work they have to do, but they lead to it in totally different ways. Development works from the less to the more perfect, and from the simpler to the more complicated, by incessant changes, small in themselves but constantly accumulating in the required direction. Design supposes that organisms were created specially on a predetermined plan, very much as the sewing-machine or self-binding reaper were constructed by their inventors.

Until quite recently all adaptations of means to ends were considered as evidences of design. A series of treatises was published some thirty years ago, for prizes left by a late Duke of Bridgewater, to illustrate this theme, among which one by Sir Charles Bell on the Hand attracted a good deal of attention. It was shown what an admirable machine the human hand is for the various purposes for which it is used, and the inference was drawn that it must have been created so by a designer who adapted means to ends in much the same way as is done by a human inventor. But more complete knowledge has dispelled this idea, and shown that the design, if there be any, must be placed very much farther back, and is in fact involved in the primitive germ from which all vertebrate life certainly, and probably all life, animal or vegetable, have been slowly developed.

The human hand is in effect the last stage of a development of the vertebrate type, or type of life in which a series of jointed vertebræ form a backbone, which protects a spinal cord containing the nervous centres, gives points of attachment for the muscles, and forms an axis of support for the looser tissues.

Certain of these vertebræ throw out bony spines or rays; at first, by a sort of simple process of vegetable growth, which formed the fins of fishes; then some of these rays dropped off and others coalesced into more complex forms, which made the rudimentary limbs of reptiles; and finally, the continued process of development fashioned them into the more perfect limbs of birds and mammals. In this last stage a vast variety of combinations was developed. Sometimes the bones of the extremities spread out, so as to form long fingers supporting the feathered wings of birds and the membranous wings of bats; sometimes they coalesced into the solid limbs supporting the bodies of large animals, as in the case of the horse; and finally, at the end of the series, they formed that marvellous instrument, the hand, as it appears in the allied genera of monkeys, apes, and man.

Any theory of secondary design and special miraculous creation must evidently account for all the intermediate forms as well as for the final result. We must suppose not one but many thousands of special creations, at a vast variety of places and over a vast extent of time; we must take into account not the successes only, but the failures, where organs appear in a rudimentary form which are perfectly useless, or in some cases even injurious, to the creature in which they are found. For instance, in the case of the so-called wingless birds, like the dodo of the Mauritius, and the apteryx of New Zealand, which are found in oceanic islands, evolution accounts readily for the atrophy or want of development of organs which were not wanted where the birds had no natural enemies and found their food on the ground; but why should they have been created with rudimentary wings, useless while they remained isolated, and insufficient to prevent their extermination as soon as man, or any other natural enemy, reached the islands where they had lived secure?

If we are to adopt the theory of design and special creation, we must be prepared to take Burns' poetical fancy as a scientific truth, and believe that Nature had to try its "prentice hand," and grope its way through repeated trials and failures from the less to the more perfect. Again, the theory of special creation must account not only for the higher organs and forms of life, but for the lower forms also. Are the bacteria, amœbæ, and other forms of life which the microscope shows in a drop of water all instances of a miraculous creation? And still more hard to believe, is this the origin of the whole parasitic world of life which is attached to and infests each its own peculiar form of higher life? Is the human tape-worm a product of design, or that wonderful parasite the trichinia, which oscillates between man and the pig, being capable of being born only in the muscles of the one, and of living only in the intestines of the other?

These are the sort of difficulties which have led the scientific world, I may say universally, to abandon the idea of separate special creations, and to substitute for it that which has been proved to be true of the whole inorganic world of stars, suns, planets, and all forms of matter; the idea of an original creation (whatever creation may mean and behind which we cannot go) of ultimate atoms or germs, so perfect that they carried within them all the phenomena of the universe by a necessary process of evolution.

This is the idea to which the Darwinian theory

leads up, by showing natural causes in operation which must inevitably tend to cause and to accumulate slight varieties, until they become large in amount and permanent, thus developing new races within old species, new species within old families, new families within old types, and new and complex types from old and simple ones.

The theory is up to a certain point undoubtedly true, and beyond that point in the highest degree probable, but scientific caution obliges us to add that it is still to a considerable extent a "theory," and not a "law." That is, it is not like the law of gravity, a demonstrated certainty throughout the whole universe, but a provisional law which accounts for a great number of undoubted facts, and supplies a framework into which all other similar facts, as at present ascertained, appear to fit with a probability not approached by any other theory, and which is enhanced by every fresh discovery made, and by the analogy of what we know to be the laws which regulate the whole inorganic world.

To enable us to talk of the "Darwinian law," and not of the "Darwinian theory," we require two demonstrations:

1. That living matter really can originate from inorganic matter.
2. That new species really can be formed from previously existing species.

As regards the first, we have seen that the efforts of science have hitherto failed to produce an instance of spontaneous generation, and all we can say is that it is probable that such instances have occurred in earlier ages of our planet, under conditions of light, heat, chemical action, and electricity, different from anything we can now reproduce in our laboratories. This, however, falls short of demonstration, and for the present we must be content to leave the origin of life as one of the mysteries not yet brought within the domain of law.

As regards the second point, we are farther advanced towards the possibility of proof. But here also we are met by two difficulties. If we appeal to historical evidence, we are met by the fact that a much greater time than is embraced by any historical record is almost necessarily required for the dying out of any old species and introduction of any new one, by natural selection. And if we appeal to fossil remains we are met by the imperfection of the geological record. As to this, it must be remembered that only a very small portion of the earth's surface has been explored, and of this a very small portion consists of ancient land surfaces or fresh water formations, where alone we can expect to meet with traces of the higher forms of animal life. And even these have been so imperfectly explored, that where we now meet with thousands and tens of thousands of undoubted human remains lying almost under our feet, it is only within the last thirty years that their existence has even been suspected. Cuvier, the greatest authority of the last generation, laid it down as an incontrovertible fact that neither men nor monkeys had existed in the fossil state, or in anything more ancient than the most superficial and recent deposits. We have now at least twenty specimens of fossil monkeys from one locality alone of the Miocene period, that of Pikermi, near

Athens, and many thousands of human remains, at least into the Quaternary period and contemporary with extinct animals, if not earlier. We must be content, therefore, with approximate solutions pointing up to but not absolutely demonstrating the truth.

What is a species? Speaking generally it is an assemblage of individuals who maintain a separate family type by breeding freely among themselves, and refusing to breed with other species. There can be no doubt that this represents what, at the first view and for a limited range of time, is in accordance with actual facts. The animal and vegetable worlds are practically mapped out into distinct species, and do not present the mass of confusion which would result from indiscriminate cross-breeding. It is clear also that this state of things has lasted for a considerable time, for the paintings on Egyptian tombs and monuments carry us back more than 4,000 years, and show us the most strongly marked varieties of the human race, such as the Semitic, the Egyptian, and the Negro, existing just as they do at the present day. They show us also such extreme varieties of the dog species as the greyhound and the turnspit, then in existence; and the skeletons of animals such as the ox, cat, and crocodile, which have been preserved as mummies, show no appreciable difference from those of their modern descendants.

When we come to look closely, however, into the matter, our faith in this absolute rule of the entire independence of species is greatly modified. In the lower grades of life we see everywhere species shading off into one another by insensible gradations, and every extension of our knowledge, both of the existing animal, vegetable, and microscopic worlds, and of those of past geological periods, multiplies instances of intermediate forms, differing from one another far less than do many of the individual varieties of recognised species. In the case of sponges, for instance, the latest conclusion of scientific research is this: that if you rely on minute distinctions as constituting distinct species, there are at least 300 species of one family of sponges, while if you disregard slight differences, which graduate into one another, and are found partly in one and partly in another variety, you must designate them all as forming only one species. Even in higher grades, as species are multiplied, it becomes more and more difficult to say where one ends and the other begins. Take the familiar instance of the grouse and ptarmigan. The red grouse is believed to be peculiar to the British Islands, while the ptarmigan is a very widely spread inhabitant of Arctic regions and high mountains. Which is more probable—that the grouse was specially created in the British Islands, apparently for the final cause of bringing sessions of Parliament to wind up business in August, or that, as the rigour of the Glacial period abated, and heather began to grow, certain ptarmigan by degrees modified their habits and took to feeding on heather tops instead of lichens, and by so doing gradually became larger birds and assumed the colour best adapted for protection in their new habitation? In point of fact, grouse showing traces of this descent in smaller size and much whiter plumage are still to be met with. It would be easy to multiply instances, but this consideration seems conclusive.

If we reject the Darwinian theory and adopt that of independent species descended from a specially created ancestor or pair of ancestors, we are driven by each discovery of intermediate or slightly modified forms, into the assumption of more and more special acts of creation, until the number breaks down under its own weight, and belief becomes impossible.

For instance, in the Madeira Islands alone, 134 species of air-breathing land-snails have been discovered by naturalists, of which twenty-one only are found in Africa or Europe, and 113 are peculiar to this small group of islands, where they are mostly confined to narrow districts and single valleys. Are we to suppose that each of these 113 species was separately created? Is it not almost certain that they are the modified descendants of the twenty-one species which had found their way there in a former geological period, when Madeira was united to Africa and Spain?

There remains only the argument from the fertility of species *inter se*, and their refusal to breed with other species. This also, when closely examined, appears to be a *primâ facie* deduction, rather than an absolute law. Different species do, in fact, often breed together, as is seen in the familiar instance of the horse and ass. It is true that in this case the mule is sterile and no new race is established. But this rule is not universal, and quite recently one new hybrid race, that of the leporine, or hare-rabbit, has been created, which is perfectly fertile. The progeny of dog and wolf has also been proved to be perfectly fertile during the four generations for which the experiment was continued. In the case of cultivated plants and domestic animals, there can be little doubt that new races, which breed true and are perfectly fertile, have been created within recent times from distinct wild species. The Esquimaux dog is so like the Arctic wolf that there can be little doubt he is either a direct descendant, or that both are descendants from a common stock. The same is true of the jackal and some breeds of dogs in the East and Africa, and other races of dogs are closely akin to foxes. But all dogs breed freely together, and can with difficulty be mated with the wild species which they so closely resemble. The modern Swiss cattle are pronounced by Rutimeyer to show undoubted marks of descent from three distinct species of fossil oxen, the *Bos primigenius*, *Bos longifrons*, and *Bos frontosus*.

There is now in the Zoological Gardens in Regent's Park a hybrid cow, whose sire was an American bison and its mother a hybrid between a zebu and a gayal. This animal is perfectly fertile, and has bred again to the bison; but what is singular is, that this hybrid resembles much more an ordinary domestic English cow than it does any of its progenitors. It is totally unlike the bison, both in appearance and disposition, and, except in having a projecting ridge over the withers, it might be mistaken for a coarse, bony, common cow. If a hybrid bull had been born of the same type, and mated with this hybrid cow, there is little doubt that a new race might have been established, extremely different from its ancestors.

In fact, nearly all the domesticated animals have the essential characters of new races. We cannot point to wild progenitors existing in any part of the world from which they are descended, and when they run wild they do not revert to any common ancestral form.

In the vegetable world instances of fertile hybrids are still more abundant, and the introduction and establishment of new varieties is a matter of every-day occurrence.

Now, whatever artificial selection can do in a short time, natural selection can certainly do in a longer time, and nothing short of absolute proof of the impossibility of species coming into existence by natural laws should induce us to fall back on the supernatural theory, with all its enormous difficulties of an innumerable multitude of special creations, most of them obviously imperfect and tentative—or rather, useless and senseless on any supposition except that of a necessary and progressive evolution. In fact, if it were not for its bearing on the nature and origin of man, few would be found to maintain the theory of miraculous creations, or to doubt that the world of life is regulated by fixed laws as well as the world of matter. But whatever touches man touches us closely, and brings into play a host of cherished aspirations and beliefs, which are too powerful to be displaced readily by calm, scientific reasoning. Shall man, who, we are told, was created in God's image and only "a little lower than the angels," be degraded into relationship with the brutes, and shown to be only the last development of an animal type which, in the case of apes and monkeys, approaches singularly near to him in physical structure? Are the saints and heroes whom we revere, and the beautiful women whom we admire, descended, not from an all-glorious Adam and all-lovely Eve, as portrayed in Milton's "Paradise Lost," but from Palæolithic savages, more rude and bestial than the lowest tribe of Bushmen or Australians? Is the account of man's creation and fall in the Hebrew Scriptures as pure a myth as that of Noah's ark, or of Deucalion and Pyrrha?

The only answer to these questions is that truth is

truth, and fact is fact, and that it is always better to act and to believe in conformity with truth and fact, than to indulge in illusions. There are many things in Nature which jar on our feelings and seem harsh and disagreeable, but yet are hard facts, which we have to recognise and make the best of. Childhood does not pass into manhood without exchanging much that is innocent and attractive for much that is stern and prosaic. Death, with its prodigal waste of immature life, its sudden extinction of mature life in the plenitude of its powers, its heart-rending separations from loved objects, is a most disagreeable fact. But it would not improve matters to keep grown-up lads in nurseries for fear of their meeting with accidents, or becoming hardened by contact with the world. Progress, not happiness, is the law of the world; and to improve himself and others by constant struggles upwards is the true destiny of man.

In working out this destiny the fearless recognition of truth is essential. Facts are the spokes of the ladder by which we climb from earth to heaven, and any individual, nation, or religion, which, from laziness or prejudice, refuses to recognise fresh facts, has ceased to climb and will end by falling asleep and dropping to a lower level.

"Prove everything, hold fast that which is true," is the maxim which has raised mankind from savagery to civilisation, and which we must be prepared to act upon at all hazards and at all sacrifices, if we wish to retain that civilisation unimpaired and to extend it further.

CHAPTER V.

ANTIQUITY OF MAN.

Belief in Man's Recent Origin—Boucher de Perthes' Discoveries—Confirmed by Prestwich—Nature of Implements—Celts, Scrapers, and Flakes—Human Remains in River Drifts—Great Antiquity—Implements from Drift at Bournemouth—Bone Caves—Kent's Cavern—Victoria, Gower, and other Caves—Caves of France and Belgium—Ages of Cave Bear, Mammoth, and Reindeer—Artistic Race—Drawings of Mammoth, etc.—Human Types—Neanderthal, Cro-Magnon, Furfooz, etc.—Attempts to fix Dates—History—Bronze Age—Neolithic—Danish Kitchen-middens—Swiss Lake-Dwellings—Glacial Period—Traces of Ice—Causes of Glaciers—Croll's Theory—Gulf Stream—Dates of Glacial Period—Rise and Submergence of Land—Tertiary Man—Eocene Period—Miocene—Evidence for Pliocene and Miocene Man—Conclusions as to Antiquity.

GREAT as the effect has been of the wonderful discoveries of modern science of which I have attempted to give a general view in the preceding chapters, there remains one which has had the greatest effect of all in changing the whole current of modern thought, viz., the discovery of the enormous antiquity of man upon earth, and his slow progress upwards from the rudest savagery to intelligence, morality, and civilisation. It is needless to point out in what flagrant and direct opposition this stands to the theory that man is of recent miraculous creation, and that he was originally endowed with a glorious nature and high faculties, which were partially forfeited by an act of disobedience. It is important, therefore, to understand clearly the evidence upon which a conclusion rests, so startling and unexpected as that which traces the origin of man back into the remote periods of geological time.

It had been long known that a stone period preceded the use of metals. Flint arrow-heads, stone axes, knives, and chisels, rude pottery, and other human remains lie scattered almost everywhere, on or near the existing surface, and are found in the sepulchral mounds and monuments which abound in all countries until they are destroyed by the progress of agriculture. These are certainly ancient, for their origin was so completely forgotten that the stone hatchets or celts (from the Latin *celtis*, or chisel) were universally believed to be thunderbolts which had fallen from heaven. But there was no proof that they were very ancient, they were always found at or near the present surface, and if animal remains were associated with them, they were those of the dog, ox, sheep, red deer, and other wild and domestic species now found in the same district. Historical record was not supposed to extend beyond the 4,000 or 5,000 years assigned to it by Bible chronology, and it was thought that this might be sufficient to account for all the changes which had occurred since man first became an inhabitant of the earth. Above all, the negative evidence was relied on, that geologists had explored far and wide, and although they had found fossil remains which enabled them to restore the characteristic fauna of so many different formations, they had found no trace of man or his works anywhere below the present surface. This seemed so conclusive that Cuvier, the greatest authority of the day, pronounced an emphatic verdict that man had not existed contemporaneously with any of the extinct animals, and probably not for more than 5,000 or 6,000 years. Here, then, appeared to be an edifice based on scientific fact, in which geologists and theologians could dwell together comfortably, and the weight of their united authority was sufficient to silence all objections, and ignore or explain away the instances which occasionally cropped up, of human remains found in situations implying greater antiquity.

Suddenly, I may almost say in a single day, this edifice collapsed like a house of cards, and the fact became apparent that the duration of human life on the earth must be measured by periods of tens, if not of hundreds of thousands of years.

It happened thus: A retired French physician, Monsieur Boucher de Perthes, residing at Abbeville, in the valley of the Somme, had a hobby for antiquarianism as decided as that of Monkbarns himself. Abbeville afforded him a capital collecting-ground for the indulgence of his tastes, as the sluggish Somme flows through a series of peat mosses, which are extensively worked for fuel, and afford many remains of the Gallo-Roman and pre-Roman or Celtic period. Higher up, on the slopes of the low hills which bound the wide valley, are numerous beds of gravel, sand, and brick-earth, which are also extensively worked for road and building materials. In these pits remains of the mammoth, rhinoceros, and other extinct animals are frequently found, and the workmen had noticed occasionally certain curiously-shaped flints, to which they gave the name of "langues du chat," or cats' tongues. Some of these were taken to Monsieur Boucher de Perthes as curiosities for his museum, and he at once recognised them as showing marks of human workmanship. This put him on the trace, and in the year 1841 he himself

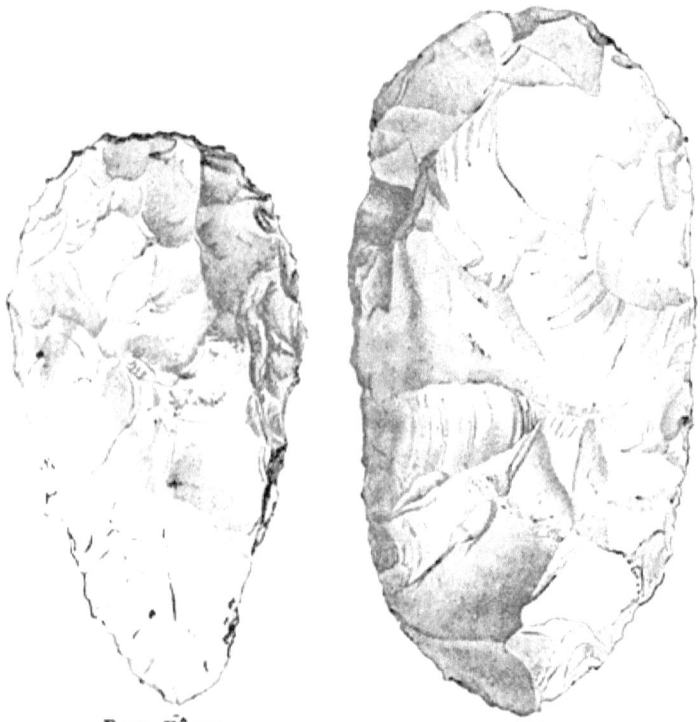

FLINT HÂCHE,
From Moulin Quignon, Abbeville.
(Half the actual size.)

FLINT HÂCHE,
From St. Acheul, Valley of the Somme.
(Half the actual size.)

(From Lubbock's "Prehistoric Times.")

discovered, *in situ*, in a seam of sand containing remains of the mammoth, a flint rudely but unmistakably fashioned by human hands into a cutting instrument. During the next few years a large quantity of gravel was removed to form the Champ de Mars at Abbeville, and many of these celts or hatchets were found. In

1847, M. Boucher de Perthes published his "Antiquités Celtiques et Antédiluviennes," giving an account of these discoveries, but no one would listen to him. The united authority of theologians and geologists opposed an infallible veto on the reception of such ideas, and it must be admitted that M. Boucher de Perthes himself did his best to discredit his own discoveries by associating them with visionary speculations about successive deluges and creations of pre-Adamite men. At length Dr. Falconer, the well-known palæontologist, who had brought to light so many wonderful fossil remains from the Sewalik hills in India, happened to be passing through Abbeville and visited M. Boucher de Perthes' collection. He was so much struck by what he saw that on arriving in London he spoke to Mr. Prestwich, the first living authority on the tertiary and quaternary strata, and Mr. Evans, whose authority was equally great on everything relating to the stone implements found in such numbers in the more recent or Neolithic period. He urged them to go to Abbeville and examine for themselves whether there was anything in these alleged discoveries. They did so, and the result was that on their return to England Mr. Prestwich read a paper to the Royal Society on the 19th May, 1859, which conclusively and for ever established the fact that flint implements of unmistakable human workmanship had been found, associated with the remains of extinct species, in beds of the Quaternary period deposited at a time when the Somme ran at a level more than 100 feet higher than at present, and was only beginning to excavate its valley.

The spell once broken evidence poured in from all quarters, and although twenty-five years only have elapsed since Mr. Prestwich's paper was read, the number of stone and other implements worked by man, deposited in museums, is already counted by tens of thousands, and they have been found from Devonshire to India, in France, England, Germany, Spain, Italy, Greece, Northern Africa, Palestine, and Hindostan, and in fact wherever they have been looked for, except in northern countries which were buried under ice during the Glacial period. Some idea of the immense number of these rude implements may be formed from the fact that the valley system of one small river, the Little Ouse, which rises near Thetford and flows into the Wash after a course of twenty-five miles, has within little more than ten years yielded about 7,000 specimens.

FLINT HACHE,
From Hoxne, Suffolk.
(Half the actual size.)
(From Lubbock's "Prehistoric Times.")

They have been found in great abundance in the valley gravels of the Thames, Ouse, Wiltshire Avon, and in fact in all the river gravels and brick-earths of the south and south-east of England; and in those of the Somme, Oise, Seine, Loire, and all the principal river systems of France; and in less numbers, probably because they have been less looked for, in similar situations over an area extending from Central and Southern Europe to Madras and China. It is a remarkable fact about these river-drift implements that they are all nearly of the same type and found under similar circumstances, that is to say, in the gravels, sands, brick-earths, and fine silt or loess deposited by rivers which have either ceased to run, or which ran at levels higher than their present ones and were only beginning to excavate their present valleys. Also they are always found in association with remains of what is known as the quaternary, as distinguished from the recent or existing fauna, and which is characterised by the mammoth, the thick-nosed rhinoceros, and other well-known types of extinct animals. The general character of these implements is very rude, implying a social condition at least as low as that of the Australian savages of the present day. They consist mainly of the flake; the chopper, or pebble roughly chipped to an edge on one side; the scraper, used probably for preparing skins; pointed flints used for boring; and by far the most abundant and characteristic of all, the *hâche* or celt, a sharp or oval implement, roughly chipped from flint or, in its absence, from any of the hard stones of the district, such as chert or quartzite, and intended to be held in the hand and used without any haft or handle.

These *hâches* are evidently the first rude type of human tools, from which the later forms of the axe, adze, chisel, wedge, etc., have been derived by a very slow and lengthened process of evolution. They differ, however, in many essential respects, from the more perfect stone celts of later periods and of modern savages. The chipping is very rude, they are never ground or polished, the pointed end is that intended for use, the butt-end being left blunt, showing that the *hâche* was not hafted but held in the hand; while the converse is always the case with the finely-chipped or polished stone celts and hatchets of the Neolithic period, which, in its later stages, are to all intents and purposes similar to modern implements, only made of stone instead of metal. But these Palæolithic *hâches* are only one step in advance of the rude natural stone which an intelligent orang or chimpanzee might pick up to crack a cocoa-nut with, or to grub up a root from the earth, or an insect from a rotten tree.

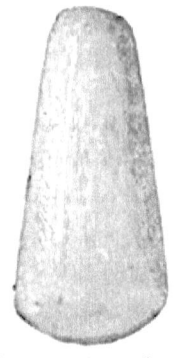

POLISHED STONE AXE.
Neolithic.
(Half the actual size.)
From Lubbock's
"Prehistoric Times."

At the same time there is not the remotest doubt as to their being the work of human hands.

 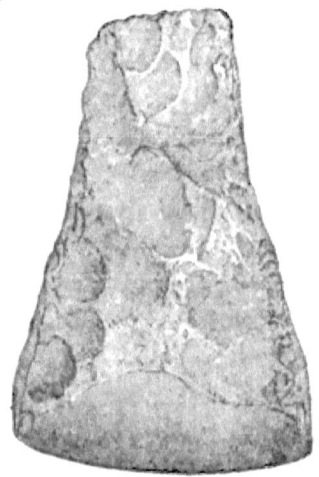

FLINT ADZE,
From Danish Kitchen-middens.

MODERN STONE ADZE,
New Zealand.

(From Lubbock's "Prehistoric Times.")

When placed side by side with the rudest forms of stone hatchets actually used by the Australian and other

DEVELOPMENT OF THE LANCE.

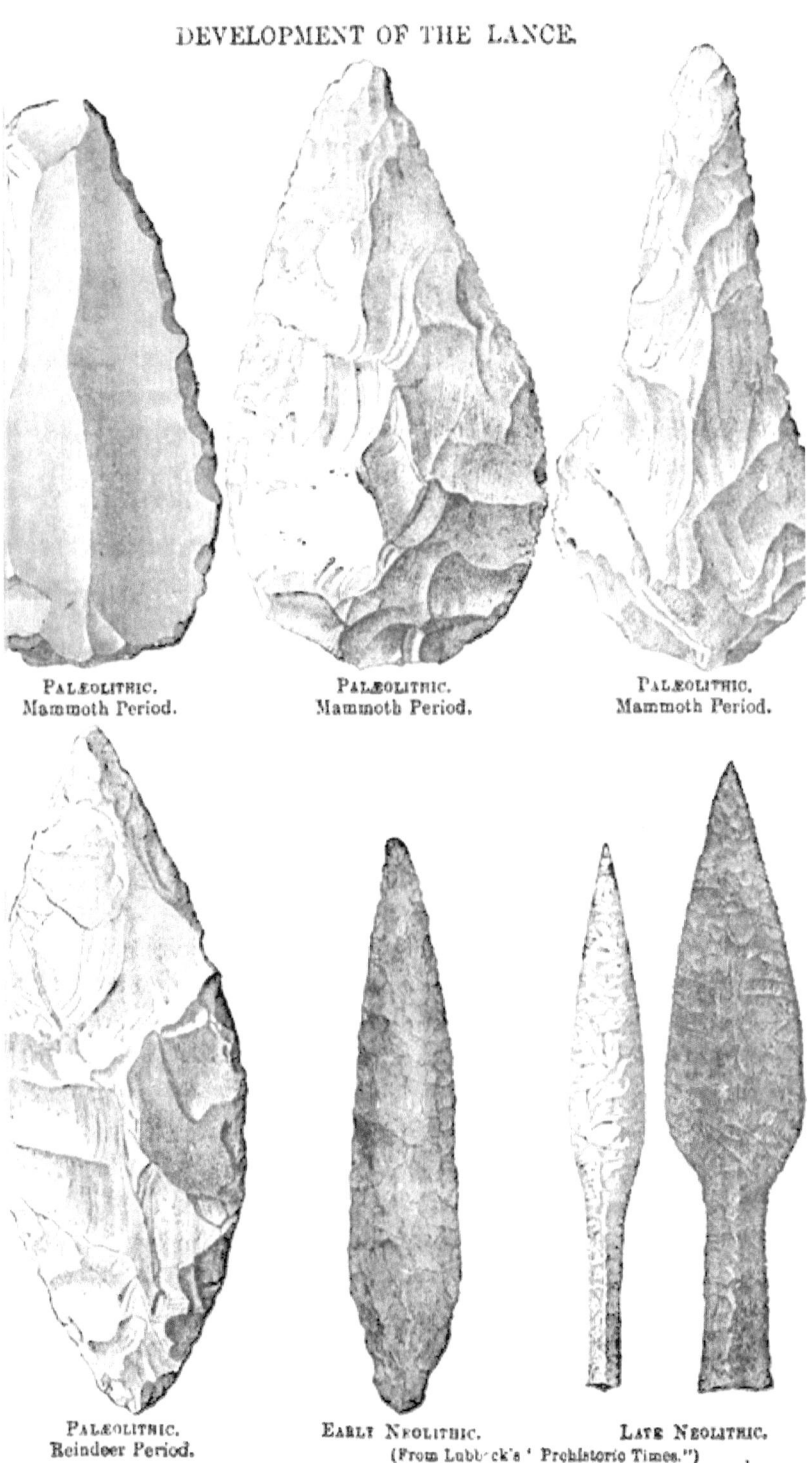

PALÆOLITHIC.
Mammoth Period.

PALÆOLITHIC.
Mammoth Period.

PALÆOLITHIC.
Mammoth Period.

PALÆOLITHIC.
Reindeer Period.

EARLY NEOLITHIC.
(From Lubbock's 'Prehistoric Times.')

LATE NEOLITHIC.

1

savages, it is difficult to detect any difference. If placed in an ascending series, from the oldest and rudest, to the finely-finished axes and arrow-heads of the period immediately preceding the use of metal, the progress may be clearly traced by insensible gradations. The blows given to bring the block to the desired shape by intentional chipping have left distinct marks; and archæologists have succeeded, with a little practice, in fashioning similar implements from modern flints. In fact, forgeries have been made by workmen in localities where collectors were eager and credulous, though fortunately such forgeries are easily distinguished from genuine antiques by the different appearance of the old and recent fractures, and other signs which make it almost impossible to deceive an experienced eye. The conclusion, therefore, of one of our best archæologists may be safely accepted, that it is as impossible to doubt that these rude stone flakes and hatchets are works of human art, as it would be if we had found clasp-knives and carpenters' adzes.

The remains of human skeletons are, as might be expected, very rare in these river drifts, which have been formed under conditions where the preservation of such remains would be very unlikely. In fact, as Sir John Lubbock points out, the bones found in the river gravels are almost invariably those of animals larger than man, such as the mammoth and rhinoceros. Still a few human bones have been found, sufficient to show that these river-drift men were probably a dolichocephalic or long and narrow-headed race, with prominent jaws, massive bones, and great muscular strength, but still, although rude and savage, of an essentially human

type, and going a very little way towards bridging over
the gap between the savage and the ape.

A more complete view, however, of the conditions
of human life at these remote periods is afforded by the
evidence given by caves, where naturally the remains
of man are much more abundant and much better
preserved. Before entering, however, on the examination of this class of evidence, it may be well to give an
instance which may help to familiarise the imagination
with the vast periods of time which must have elapsed
since Palæolithic man left these rude implements
within reach of river floods.

Among the gravels in which Palæolithic *hâches* have
been found, are some which cap the cliff at Bournemouth at a height of about 130 feet above the sea.
This gravel can be traced in a gradual fall from west to
east, along the Hampshire coast and the shores of the
Solent to beyond Spithead, and was evidently deposited by a river which carried the drainage of
the Dorsetshire and Hampshire downs into the sea
to the eastward, and of which the present Avon,
Test, and Itchen were tributaries. But for such a river
to run in such a course the whole of Poole and Christchurch bays must have been dry land, and the range
of chalk downs now broken through at the Needles
must have been continuous. To borrow the words of
Evans in his "Ancient Stone Implements," "Who,
standing on the edge of the lofty cliff at Bournemouth,
and gazing over the wide expanse of waters between
the present shore and a line connecting the Needles
on the one hand and the Ballard Down Foreland on
the other, can fully comprehend how immensely remote

was the epoch when what is now that vast bay was high and dry land, and a long range of chalk downs, 600 feet above the sea, bounded the horizon on the south? And yet this must have been the sight that met the eyes of those primeval men who frequented the banks of that ancient river which buried their handiworks in gravels that now cap the cliffs, and of the course of which so strange but indubitable a memorial subsists in what has now become the Solent Sea."

Any attempt to assign a more precise date than the vague one of immense antiquity to these early traces of primeval man, had better be postponed until we have examined the more detailed and extensive body of evidence which has been afforded by the exploration of caves, to which the great discovery at Abbeville at once gave an immense impulse, and which has since been prosecuted in England, France, Belgium, and Germany, with the greatest ardour and success.

The caves in which fossil remains are found occur principally in limestone districts. They are due to the property which water possesses, when charged with a small quantity of carbonic acid, of dissolving lime. Rain falling on the earth's surface takes up carbonic acid from contact with vegetable matter, and a portion of it finds its way through cracks and crevices in the subjacent rock to lower levels, where it comes out in springs of hard water charged with carbonate of lime from the rock which it has dissolved. It has been calculated that the average rainfall on a square mile of chalk thus carries away about 140 tons of solid matter in a year. In this way underground channels are formed, some of which become large enough to admit of streams flowing through them, and even rivers, as is seen in the limestone district of Carinthia, where considerable rivers are swallowed up and run for miles beneath the surface. In this way caverns are formed, or sometimes a series of caverns, which represent the pools of the rivers which formerly flowed through them. Accumulations were formed at the bottom of these pools of whatever may have been brought down by the stream, and when, owing to changes in level or denudation of the gathering grounds, the rivers ceased to flow in the old channel, these pools became dry and were converted into caves, in which wild beasts and man found shelter and left their remains. The *débris* thus formed accumulated with a mixture of blocks which fell from the roof, and of red loamy earth consisting of the residue of the limestone rock insoluble in water, and of dust and mud brought in by winds and floods, and occasionally interstratified by beds of stalagmite, composed of thin films of crystalline carbonate of lime, deposited drop by drop by drippings through the rock forming the roof of the cave. These drippings form what are called stalactites, which hang like pendent icicles from the roof of caves, and as the drip falls from these it forms a corresponding deposit, known as stalagmite, on the floor below. The formation of this deposit is necessarily extremely slow, and it only goes on when the drops of water charged with a minute excess of carbonate of lime come in contact with the air; so that whenever the floor of the cave was under water no stalagmite could be formed. The alternations, therefore, of deposits of stalagmite represent alternations of long periods during which the cave was generally dry or generally flooded. During the dry periods, when the cave happened to be inhabited, the treadings on the floor would prevent the accumulation of an unbroken deposit of pure stalagmite, and the crystalline matter would be employed in forming a solid cement of the various *débris* into what is known as a breccia.

Another class of caves, or rock-shelters, has been formed along the sides of valleys bounded by cliffs, where the stratification is horizontal or nearly so; but the different beds vary much in hardness and permeability to water. The softer strata weather away more rapidly than the others, and thus form shallow caves or deep recesses in the face of the cliffs, with a floor of hard rock below and a roof of hard rock above, which afford dry and commodious shelters for any sort of animal, including man. In other respects they resemble the first class of caves in having their contents cemented into a breccia by the dripping of water charged with carbonate of lime from the roof, and, if the cave happened to be deserted for a long period, this deposit would in the same way form a bed of stalagmite and seal up securely everything below it. In some cases, also, the roof would fall in, and thus preserve everything previously existing in the cave for the investigation of future geologists.

With these general remarks readers will be able to understand the evidence afforded by the remains of man found in caverns. I will begin by taking as a typical case that of Kent's Cavern, near Torquay, because it is one of the earliest and best known, and all the facts concerning it have been verified by explorations carefully conducted by a committee appointed by the British Association in 1864, and which comprised the names of the most eminent authorities in geology and palæontology, including those of Sir Charles Lyell, Sir John Lubbock, Mr. Evans, Mr. Boyd Dawkins, Mr. Pengelley, and others.

The cave is about a mile east from Torquay harbour, and runs into a hill of Devonian limestone in a winding course, expanding into large chambers connected by narrow passages. The following is the series of deposits in descending order in the large chamber near the entrance:

1. Large blocks of limestone which have fallen from the roof.
2. A layer of black, muddy mould, three inches to twelve inches thick.
3. Stalagmite one foot to three feet thick.
4. Red cave-earth with angular fragments of limestone of variable thickness, but in places five to six feet thick.

In the black earth above the stalagmite were found a number of relics of the Neolithic or polished stone period, with a few articles of bronze and pottery, some of which appear to be of a date as late as that of the Roman occupation of Britain. Associated with these are bones of ox, sheep, goat, pig, and other ordinary forms of existing species, and there is an entire absence of any older fauna, or of any of the ruder forms of Palæolithic implements. When we get below the stalagmite into the underlying cave-earth, the case is entirely reversed. Not a single specimen of polished or finely-wrought stone, or of pottery, is to be found; a vast number of celts or *hâches*, scrapers, knives, hammer stones, and other stone implements, are met with, which are all of the rude Palæolithic type found in the river drifts, with a few bone implements such as harpoon-heads, a pin, an awl, and a needle, like those frequently met with in the caves of France and Belgium. Associated with these are a vast number of bones and teeth, all of which belong to the old quaternary fauna, of which many species have become extinct and others have migrated to distant latitudes.

The following is a list of the mammalian remains which have been found in this cave-earth below the stalagmite:

ABUNDANT.

The Cave Lion, a large extinct species of lion.
Cave Hyæna,   ,,   ,,   hyæna.
Cave Bear,   ,,   ,,   bear.
Grizzly Bear.
Mammoth (*Elephas primigenius*).
Rhinoceros (*Tichorinus*), woolly or thick-nosed extinct species.
Horse.
Bison.
Irish Elk.
Red Deer.
Reindeer.

SCARCE.

Wolf.
Fox.
Glutton.
Brown Bear.
Urus.
Hare.
Lagomnys, tailless Arctic hare.
Water Vole.
Field Vole.
Bank Vole.
Beaver.
And one specimen of the Machairodus, or Great Sabre-toothed Tiger, which is one of the characteristic species of the upper Miocene and Pliocene formations.

These constitute a fauna which is characteristic of the Pleistocene, Quaternary, or Palæolithic period, and essentially different from that of the prehistoric or Neolithic period, which is practically the same as that now existing. Wherever remains of the mammoth, woolly rhinoceros, and cave bear are found, Palæolithic implements may be expected, and conversely. In fact Palæolithic man is as essentially part of the characteristic fauna of the Quaternary period, as the Palæotherium is of the Eocene, or the Deinotherium and Hipparion of the Miocene.

A large number of other caves have been explored in England, notably the Victoria Cave near Settle in Yorkshire, the Gower Caves in South Wales, the Brixham Cave in Devonshire, the Woking Cave in Somersetshire, and King Arthur's Cave in Herefordshire, and the results have been everywhere practically the same as those at Kent's Cavern. The same class of implements have been found and the same fauna, with the occasional addition of a few species, among which the hippopotamus is the most remarkable. Everywhere there is the same entire break between the Neolithic and the Palæolithic deposits, and the same evidence of great antiquity for the latter. It would appear as if in the British area some great geological change, such as submergence beneath the sea or invasion of the ice, had exterminated or driven away Palæolithic man, along with the mammoth, rhinoceros, cave bear, and other extinct animals of the Palæolithic fauna, and after a long lapse of time the area had again become habitable and been occupied by a newer race and by the recent fauna.

The same remark applies to the river drifts, which

not in England only, but everywhere, appear to belong to a distinct period, vastly more ancient than any of the recent deposits in which Neolithic remains are found. So far, therefore, as the river drifts and British caves are concerned, all that we could say of the Palæolithic period is that it is of vast antiquity, and must have lasted for an immense time, as it was in force for the whole time requisite for rivers like the Somme or Avon, which drain small areas, to cut down their present valleys, often two or three miles wide, from the level of their upper gravels, which are in many places 100 to 150 feet above the level of the highest floods of the present rivers.

But the caves of France and Belgium supply us with more evidence, and enable us to trace the history of long periods of Palæolithic time, and study in detail the succession of changes that have occurred, and the habits, arts, and industries of the various tribes of primitive men who occupied these caves and rock-shelters at these remote periods. In fact, it may be said with truth that we know more about the men who chased the mammoth and reindeer in the South of France perhaps 50,000 years ago, than we do about those who lived there immediately before the classical era, or less than 5,000 years ago.

In certain provinces of France and Belgium it happens fortunately that there are extensive districts of limestone, in which caverns and rock-shelters are extremely abundant and full of Palæolithic remains in an excellent state of preservation. The abundance of such caves may be estimated from the fact that the cliffs, bounding one small river, the Vezère, in the department of Dordogne in the South of France, contain in a distance of eight or ten miles no fewer than nine different stations, each of which has given a vast variety of remains embedded in the breccias and cave-earths of their respective floors; and the small river Lesse in Belgium has been scarcely less prolific. Of the abundance of the human and animal remains found in such caverns it may be sufficient to say that one alone, that of Chaleux in the valley of the Lesse, is computed by Dumont to have yielded not less than 40,000 distinct objects.

The great abundance of remains thus collected, both of human bones and implements, and of animals contemporaneous with them, have made it possible to classify and arrange, in relative order of time, a good many of the subdivisions of the Palæolithic period. This has been done partly by the order of superposition and partly by the greater or less rudeness of the implements of stone and bone, and by the greater or less abundance of those animals of the quaternary fauna which appeared first and disappeared soonest. The result has been to show that the period when vast herds of reindeer roamed over the plains of Southern France up to the Pyrenees was not the earliest, but was preceded by a long period when the reindeer was scarce, and the remains of the mammoth, cave bear, and cave hyæna were more abundant than in the following ages. The implements of this period are of the earlier river-drift type and extremely rude, and there is an almost entire absence of instruments of bone.

Gradually as we pass upwards the more Southern forms of elephant, rhinoceros, antelopes, and great carnivora disappear, and the mammoth and cave bear become scarcer, while the reindeer becomes more and more abundant until at length it furnishes the chief source of food, and its horns one of the principal materials for the manufacture of implements. Concurrently with this change we find a progressive improvement in the arts of life, as shown by stone implements more carefully chipped into a greater variety of forms, and arrow and lance-heads, barbed harpoons, awls, and needles for sewing skins, made chiefly from the antlers of the reindeer.

At length we arrive at one of the most interesting facts disclosed by these researches, that during one of the later or reindeer periods of the Palæolithic era, many of the caves in the South of France, and also in Switzerland and Southern Germany, were occupied by a race who, like the Esquimaux of the present day, had a strong artistic tendency, and were constantly drawing with the point of a flint on stone or bone, or modelling with flint knives from horns and bones, sketches of the animals they hunted, scenes of the chase, or other objects which struck their fancy. These are exceedingly well done, so that there is no difficulty in recognising the animals intended to be represented, among which are the mammoth, cave bear, reindeer, wild horse, and wild ox. The sketch of the mammoth which is engraved on a piece of ivory, from the cave of La Madeleine in the valley of the Vezère, is particularly interesting, as it corresponds exactly with the mammoth whose body was found entire in frozen mud on the banks of a river in Siberia, and it sets at rest all possible question of man having been really contemporary with this extinct animal in the South of France.

The drawings and carvings of other animals,

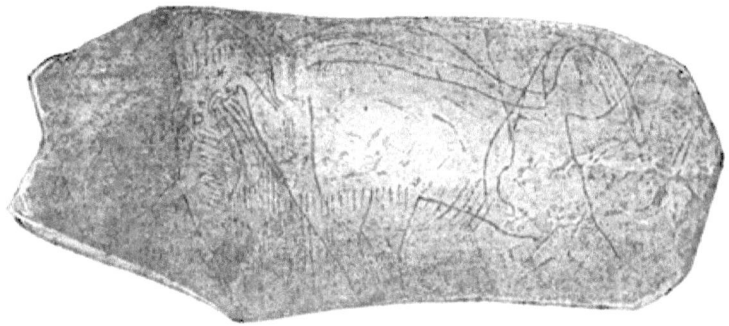

PORTRAIT OF MAMMOTH.
Drawn with a flint on a piece of Mammoth's ivory; from Cave of La Madeleine, Dordogne, France.

EARLIEST PORTRAIT OF A MAN, WITH SERPENT AND HORSES' HEADS.
From Grotto of Les Eyzies.   Reindeer Period.

REINDEER FEEDING.
From Grotto of Thayngen, near Schaffhausen, Switzerland.

especially of the reindeer, are often extremely spirited, and one especially of a reindeer engraved on a bit of bone from a cave at Thayngen, near Schaffhausen in Switzerland, would do credit to any modern animal painter. A very few human figures are found among these primeval drawings, but strangely, while the animals are so well drawn, those of men are very inferior and almost infantine in execution. They are sufficient, however, to show that the savage of Périgord pursued the formidable aurochs, naked, armed with a lance or javelin, bearded on the chin but not on the rest of the face, and wearing his hair in a tuft on the top of the head.

We do not, however, depend on these drawings for evidence of the sort of men who inhabited these caves in Palæolithic days. A large number of skulls and complete skeletons have been found in different caves, some of which have served as sepulchral vaults for families and tribes, while in others individuals have been crushed by falls of rock, or otherwise interred, and in a few cases skulls and bones have been found at great depths in river drifts, and in the loess, or fine glacial mud which fills up the valley of the Rhine and other areas over which the great Swiss glaciers when melting poured their turbid streams.

The most celebrated of these are :

The Neanderthal and Canstadt skulls, which are considered to belong to the oldest type, having been found in the lowest strata, which contain the rudest implements and the most archaic fauna. Of these the Neanderthal skull has attracted much attention from its singularly brutal appearance, having a very low and receding forehead, and a massive bony ridge over the eyes resembling

ANTIQUITY OF MAN. 127

that of the gorilla. But the brain is of fair capacity, and occasional skulls of a similar type occur at the present day, so that we are not warranted in saying that we have discovered the "missing link" between man and ape, especially as the Engis and other skulls of this period present less exceptional features. All we can safely say is that the oldest type of man known to us seems to have been characterised by long and narrow heads, prominent eyebrows, medium stature,

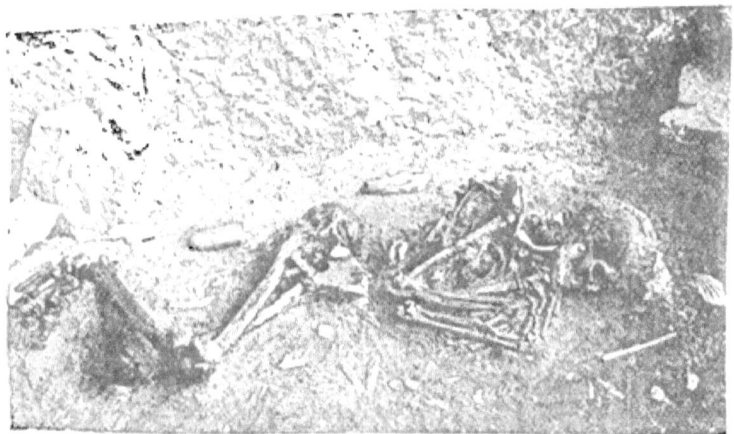

MENTONE SKELETON. Palæolithic. Reindeer Period.

and great thickness of bones and prominence of ridges denoting great muscular strength.

The discovery of a sepulchral chamber at Cro-Magnon in the valley of the Vezère, with several entire skeletons, gave evidence of another type which has been found elsewhere in caves of the same age, viz., newer than the earliest mammoth and cave bear age to which the oldest skulls are referred, but older than the subsequent reindeer age, and still characterised by great rudeness of implements. This is a remarkable type, for these savages were really a fine race of men, tall in stature, and with well-developed brain. They are long-headed, but not more so than is often found in the best modern European skulls, and the average capacity of the skull exceeded that of most modern races, while their average height was not less than 5 ft. 10 in. for the men, and 5 ft. 6 in. for the women.

Another totally different race appears in caves of the same period or a little later, which is known as the Furfooz race, from a sepulchral cave in Belgium where a number of skeletons were discovered, but which appears to have been widely spread throughout Europe towards the middle of the Palæolithic period. The type of this race is almost exactly that of the modern Lapp, short in stature, averaging not above 5 ft., though strong and muscular, and with small round heads and high cheek-bones. From this time forward, long and short-headed races, and intermediate types resulting probably from their intermixture, seem to have existed pretty much as they do at the present day, and the important conclusion to be drawn is, that even as far back as the early Glacial period, man had already existed long enough to develop different races, and in sufficient numbers to scatter wandering tribes of savage hunters widely over the earth and up to the verge of glaciers and the utmost confines of inhospitable regions.

In trying to fix anything like definite dates for man's existence upon earth, we must reverse the process by which we have proved the enormous antiquity of his earliest remains, and ascend step by step from the known to the unknown. The first step is that supplied by history.

Authentic Egyptian history begins with Menes, the

first king who united the different provinces of Egypt into one empire.

The date of this event has been fixed by the best authorities, who have devoted their lives to the study of Egyptian texts and monuments, at about 5,000 years B.C., or say 7,000 years before the present time. Boeck makes it B.C. 5702, Unger 5613, Mariette 5004, Brugsch 4455, Lauth 4157, Lepsius 3892, and Bunsen 3623.

It will be observed that the tendency of all the more recent investigations is to lengthen the date, and that of Mariette may be safely assumed as the *minimum* limit of time for the foundation of the Egyptian monarchy.

Now this date shows no trace of approach to a primitive and uncivilised state of things. On the contrary, Menes is related to have carried out a great engineering work by which the Nile was embanked, its course changed, and the new capital city of Memphis built on the site reclaimed. His next successor, Tet, is credited with having written learned treatises on medicine and anatomy, and the earliest pyramid, that of Sakkara, was probably built by a king who ascended the throne only eighty-eight years after the death of Menes.

The annals and monuments of Chaldæa and China take us back to about 2,500 years B.C., or say for 4,500 years from the present time, and tell the same tale as those of Egypt of dense population and a high degree of civilisation already established. In fact, it is evident that the great alluvial valleys of rivers such as the Nile and Euphrates have been inhabited for a number of centuries by a population who had emerged from the hunter and pastoral stage into that of agriculture, and had increased and multiplied until great cities were built and mighty monarchies founded, and who were in possession of most of the arts of civilised life. The Egyptian date which carries us back about 7,000 years is, however, by far the earliest upon which we can rely as an authentic record, and any glimmerings of history beyond this are obviously mythical.

Here, then, we take leave of history, and must explore our way upwards by the aid of archæology and geology.

The earliest historical civilisations were all acquainted with metals, chiefly in the form of bronze, which is an alloy of copper and tin, very hard, easily cast, and well adapted for every description of tool and weapon. Indeed, it has only been superseded by iron within recent historical times. But the Bronze Age was preceded by a long Neolithic period, when stone, finely wrought and often ground or polished, was used for the purposes to which metal was afterwards applied. The men of this Neolithic period were comparatively civilised; they had all the common domestic animals, the dog, horse, ox, sheep, goat, and pig; also some of the cultivated grains, as wheat and barley; they wore clothing and lived in villages. According to all appearance they were the first wave of the great migrations into Europe from Asia, and either occupied regions left empty by the last vicissitudes of the Glacial period, or conquered, and partly exterminated and partly intermixed with, the ruder savages of the Palæolithic period. Some think the Iberian or Basque people may be a remnant of this Neolithic race, who were driven westward by the later wave of Celtic migration just as the Celts were by the still later waves of Teutonic and Slavonic immigrants. Be this as it may, it is certain that a Neolithic people were spread very widely over the globe, as their remains of very similar character are found almost everywhere in Europe, Asia, and America, and always in association with the existing or most recent fauna and configuration of the earth's surface.

The difficulty in assigning any precise date for these remains arises very much from the fact that the Neolithic passed into the Bronze or historical civilisation, at different times in different countries. The Australians, the Polynesians, and the Esquimaux were or are still in the Stone period, while steam-engines are spinning cotton at Manchester, and the most famous cities of Egypt and the East have been for centuries buried under shapeless mounds of their own ruins. It is probable that all Europe remained in the Neolithic stage for many centuries after the historical date of the commencement of the Egyptian empire.

Still there are some remains which may enable us to form an approximate conjecture of the time during which this Neolithic period may have lasted.

The two principal clues are furnished:

1. By the Danish mosses and kitchen-middens.
2. By the Swiss lake-dwellings.

In Denmark there are a number of peat mosses varying in depth from ten to thirty feet, which have been formed by the filling up of small lakes or ponds in hollows of the Glacial drift. Around the borders of these mosses, and at various depths in them, lie trunks of trees which have grown on their margin. At the present surface are found beech-trees, which are now, and have been throughout the whole historical period of 2,000 years, the prevalent form of forest vegetation in Denmark. Lower down is found a zone of oaks, a tree which is now rare and almost superseded by the beech. And still lower, towards the bottom of the mosses, the fallen trees are almost entirely Scotch firs, which have been long unknown in Denmark and when introduced will not thrive there. It is evident, therefore, that there have been three changes of climate, causing three entire changes in the forest vegetation of Denmark, since these mosses began to be formed. The latest has lasted certainly for 2,000 years and we cannot tell how much longer, so that some period of more than 6,000 years must be assumed for the three changes.

Now, it is invariably found that remains of the Iron Age are confined to the present or beech era, while bronze is found only in that of oak, and the Age of Stone coincides with that of the Scotch fir.

The kitchen-middens afford another memorial of the prehistoric age in Denmark. There are mounds found all along the sheltered sea-coasts of the mainland and islands, consisting chiefly of shells of the oyster, cockle, limpet, and other shell-fish, which have been eaten by the ancient dwellers on these coasts. Mixed up with these are the bones of various land animals, birds, and fish, and flint flakes, axes, worked bones and horns, and other implements, including rude hand-made pottery. The relics are very much the same as those found in the fir zone of the peat mosses, and although old as compared with the Iron or historical age, they do not denote any extreme antiquity. The shells are all of existing species, though the larger size of some of those found on the shores of the Baltic shows that the salt water of the North Sea had then a freer access to it than at present. The bones of animals, birds, and fish are also all of existing species, and no remains of extinct animals, such as the mammoth, or even of reindeer, have been found. By far the most common are the red deer, roe-deer, and wild boar. The dog was known, but appears to have been the only domestic animal.

Most of the stone implements are rude, but a few carefully-worked weapons have been found, and a few specimens of polished axes, which, with the presence of pottery and the nature of the fauna, show conclusively that these Danish remains are all of the Neolithic age and subsequent to the close of the Glacial period. In fact, similar shell mounds are found in almost all quarters of the globe where savage tribes have lived on the sea-coast, subsisting mainly on shell-fish, and they are probably still being formed on the shores of the Greenland and Arctic Seas, and in Australia, and remote islands of the Pacific.

Human remains are scarce in these Danish deposits, but numerous skulls and skeletons have been found in tumuli which, from their situation and from stone implements being buried with the dead, may be reasonably inferred to be those of the people of the peat mosses and shell mounds. They denote a short race with small and very round heads, in many respects resembling the present Lapps, but with a more projecting ridge over the eye.

On the whole, all we can conclude from these Danish remains is that at some period, not less than

6,000 or 7,000 years ago, when civilisation had already been long established in the valley of the Nile, rude races resembling the Lapps or Esquimaux lived on the shores of the Baltic, who, although so much more recent, and acquainted with the domestic dog, pottery, and the art of polishing stone, had not advanced much beyond the condition of the later cave-men of the South of France; and that this race was succeeded by one who brought in the much higher civilisation of the Bronze Age.

The lake-dwellings of Switzerland give still more detailed and interesting information as to Neolithic times.

During a very dry summer in 1854, the Lake of Zurich fell below its usual level and disclosed the remains of ancient piles driven into the mud, from which a number of deer-horns and other implements were dredged up. This led to further researches, and the result has been that a large number of villages built on these piles has been discovered in almost all the Swiss lakes, as well as in those of Italy and other countries. On the whole, more than 200 have been discovered in Switzerland, and fresh ones are being constantly brought to light. They range over a long period, a few belonging to the Iron and even to Roman times; while the greater number are almost equally divided between the Age of Bronze and that of Stone. Some of them are of large size, and must have been long inhabited and supported a numerous population, from the immense number of implements found, which at one station alone, that of Concise on the Lake of Neufchâtel, amounted to 25,000. These implements consist mainly of axes, knives, arrow-heads, saws, chisels, hammers, awls, and needles, with a quantity of broken pottery, spindle-whorls, sinkers for nets, and other objects.

In the oldest stations, where no trace of metal is found, and the decay of the piles to a lower level shows the greatest antiquity, the implements are all of the Neolithic type, and the animal remains associated with them are all of the recent fauna. There are no mammoths, rhinoceroses, or reindeer; the wild animals are the red deer and roe, the urus, bison, elk, bear, wolf, wild cat, fox, badger, wild boar, ibex, and other existing species; and of domestic animals, the dog, pig, horse, goat, sheep, and at least two varieties of oxen. Birds, reptiles, and fish, were all of common existing species. Carbonised ears of wheat and barley have been found, as also pears and apples, and the seeds, stones, and shells of raspberry, blackberry, wild plum, hazel-nut, and beech-nut. Twine, and bits of matting made of flax, as well as the occurrence of spindle-whorls, show that the pile-dwellers were acquainted with the art of weaving.

On the whole, these pile-villages show that a large population lived in Switzerland for a long time before the dawn of history, who had already attained a considerable amount of civilisation at their first appearance, which went on steadily increasing down to the time of the Roman conquest. Various attempts have been made to fix an approximate date for the earliest of these pile-villages, but they have not been very successful. They have been based mainly on the amount of silting up which has taken place in some of the smaller lakes since the piles were driven in, as compared with that which has occurred since the Roman period. The best calculations appear to show that 6,000 or 7,000 years ago Switzerland was already inhabited by men who used polished stone implements, but how long they had been there we have no distinct evidence to show. Perhaps 10,000 years may be taken as the outside limit of time that can be allowed for the Neolithic period in Switzerland, Denmark, or any known part of Europe.

In Egypt, however, there is evidence of a much greater antiquity. Fragments of pottery, which was entirely unknown in the Palæolithic age, have been brought up by borings in the Nile Valley from depths which, at the average rate of accumulation there during the last 3,000 years of three inches and a half in a century, would denote an age of from 13,000 to 18,000 years. Looking at the dense population and high civilisation of Egypt at the commencement of history, 7,000 years ago, it is highly probable that this time at least must have elapsed since the country was first occupied by a settled agricultural population as far advanced in the arts of life as the lake-dwellers of Switzerland.

Any calculation, however, of Neolithic time takes us back a very short step in the history of the human race. The Palæolithic period must evidently have been of vastly longer duration.

Any attempt to estimate this must depend entirely on geological considerations. Palæolithic man is part of the Quaternary fauna, which came in with the commencement and continued down to the close of the great Glacial period.

In carrying our researches further back, the possibility of assigning anything like a definite date for the existence of man depends, therefore, on the question whether it is possible to fix any approximate dates for the commencement and duration of this period.

In the first place, how do we know that there has been a Glacial period?

In England we are familiar with water, but not with ice; we therefore recognise at once the signs of the action of water. If we come across a dry channel, winding in alternating curves between eroded banks, and showing deposits of gravel and silt, we say without hesitation, "Here a river formerly ran." But if we had lived in Switzerland, we should recognise with equal certainty the signs of glacial action. Suppose any one visiting Chamouni walks up the valley to the foot of the Mer de Glace, where the Arve issues from the glacier, let us say in autumn, when the front of the glacier has shrunk back some distance, what does he see? Rounded and polished rocks, which seem as if they had been planed by a gigantic plane working downwards over them, and on these a mass of miscellaneous rubbish shot down as if from a dust-cart, consisting of stones of all sizes, some of them boulders as big as a house, scattered irregularly on a mass of clay and sand. When he looks more closely he will see that these stones are not rounded as they would be by running water, but blunted at their angles by a slow grinding action; and in many cases, both the stones and the rocks on which they rest are scratched and striated in a direction which is that of the glacier's motion. At the bottom of this rubbish-heap he will find the clay into which the rock has been ground by the full weight of the glacier, very stiff and compact;

while if he looks down the valley, he will see, on a hot day, a swollen and turbid river issuing from the melting ice and flooding the meadows, on which it will leave a deposit of fine mud. These are effects actually produced by ice; and wherever he sees them he can infer the former presence of a glacier, as certainly as when he sees a bed of rounded pebbles he infers the former presence of running water. The planed rocks are commonly known as *roches moutonnées*, from a fancied resemblance of their smooth, rounded hummocks to the backs of a flock of sheep lying down; the rubbish-heaps are called *moraines;* and the stiff bottom clay with boulders embedded in it is called the *grund-moraine*, till, or boulder clay; while the blunted and scratched stones are said to be glaciated.

These tests, therefore, *roches moutonnées*, moraines, boulders, and glaciated stones, are infallible proofs that wherever we find them there has been ice-action, either in the form of glaciers, or of icebergs, which are only detached portions of glaciers floated off when the glacier ends in the sea. Now, if our inquirer extends his view, he will find that these signs, the meaning of which he has learned at the head of the valley of Chamouni, are to be found equally in every valley and over the whole plain of Switzerland, up to a height of more than 3,000 feet on the slope of the opposite Jura range, while on the Italian side the Glacial drift extends far into the plains of Piedmont.

Extending our view still more widely, we find that every high mountain range in the Northern hemisphere has had its system of glaciers; and one great mountain mass, that of Scandinavia, has been the nucleus of an enormous ice-cap, radiating to a distance of not less than 1,000 miles, and thick enough to block up with solid ice the North Sea, the German Ocean, the Baltic, and even the Atlantic up to the 100 fathom line. This ice-cap, coalescing with local glaciers from the higher lands of England, Scotland, and Ireland, swept over their surface, regardless of minor inequalities of hill and valley, as far south as to the present Thames Valley, grinding down rocks, scattering drift and boulders, and, in fact, doing the first rough sub-soil ploughing which prepared most of our present arable fields for cultivation. The same ice-sheet spread masses of similar drift over Northern Germany, Sweden, Denmark, and the northern half of European Russia, and left behind it numerous boulders which must have travelled all the way from Norway or Lapland.

If we cross the Atlantic we find the same thing repeated on a still larger scale in North America. A still more gigantic ice-cap, radiating from the Laurentian ranges, which extend towards the Pole from Canada, has glaciated all the minor mountain ranges to the south up to heights sometimes exceeding 3,000 feet, and coalescing with vast glaciers thrown off by the Rocky Mountains from their eastern flanks, has swept over the whole continent, leaving its record in the form of drift and boulders, down to the 40th parallel of latitude. It is difficult to realise the existence of such gigantic glaciers, but the proofs they have left are incontrovertible, and we have only to look to Greenland to see similar effects actually in operation. The whole of that vast country, where at former periods of the earth's history, fruit-trees grew and a genial climate prevailed, is now buried deep under one solid ice-cap, from which only a few of the highest peaks protrude, and which discharges its surplus accumulation of winter snow by huge glaciers filling all the fiords and pushing out into the sea with an ice-wall sometimes forty or fifty miles in length, from which icebergs are continually breaking off and floating away. A still more gigantic ice-wall surrounds the Southern Pole, and in a comparatively low latitude presented an insuperable barrier to the further progress of the ships of Sir J. Ross's expedition.

A still closer examination of the Glacial period shows that it was not one single period of intense cold but a prolonged period, during which there were several alternations, the glaciers having retreated and advanced several times with comparatively mild inter-glacial periods, but finally with a tendency on each successive advance to contract its area, until the ice shrank into the recesses of high mountains, where alone we now find it. Another noteworthy point is that during this long Glacial period there were several great oscillations in the level of sea and land.

Such was the Glacial period, and to assign its date is to fix the date when we know with certainty that man already existed, and had for some long though unknown time previously been an inhabitant of earth. Is this possible? To answer this question we must begin by considering what are the causes, or combination of causes, which may have given rise to such a Glacial period. When we look at the causes which actually produce existing glaciers, we find that extreme cold alone is not sufficient. In the coldest known region of the earth, in Eastern Siberia, there are no glaciers, for the land is low and level and the air dry. On the other hand, in New Zealand, in the latitude of England and with a mean annual temperature very similar to that of the West of Scotland, enormous glaciers descend to within 700 feet of the sea-level. The reason is obvious; the Alps of the South Island rise to the height of 11,000 feet above the sea, and the prevalent westerly winds strike on them laden with moisture from their passage over a wide expanse of ocean. In like manner, in the case of the Swiss Alps, the Himalayas, and other great mountain ranges, high land and moist winds everywhere make glaciers. Given the moist wind, any great depression of temperature, whether arising from elevation of land or other causes, will make it deposit its moisture in the form of snow, and the accumulation of snow on a large surface of elevated land must inevitably relieve itself by pushing down rivers of ice to the point where it melts, just as the rain-fall relieves itself by pouring down rivers to the point where the surplus water finds its level in the sea.

When the two conditions of high land and moist winds are combined, low temperature increases their effect, and the snow-fall consolidates into a great ice-cap, from which only the tops of the highest mountains project, and which pushes out gigantic glaciers far over surrounding countries and into adjacent seas. Such is now the case in Greenland, and was formerly the case in Scandinavia, where a huge sheet of ice radiated from it over Northern Germany as far as Dresden, filled up the North Sea, and, coalescing with smaller ice-caps from the highlands of Scotland, England, and Wales, buried the British Islands up to the Thames under massive ice. At the same period glaciers from the

Alps filled the whole plain of Switzerland, and in North America the ice-cap extended from Labrador to Philadelphia.

The first remark to be made is that, as these phenomena depend primarily on moist winds, and only secondarily on cold, and as moist winds imply great evaporation and therefore great solar heat over extensive surfaces of water, all explanations are worthless which suppose a general prevalence of cold, either from less solar radiation, passage through a colder region of space, or otherwise. We must seek for a cause which is consistent with the general laws of Nature, and with the leading facts of the actual generation of glaciers at the present day.

Astronomers believe that they have discovered such a cause, in the theory first started by Mr. Croll, that the glaciation of the Northern hemisphere was due to a secular change in the shape of the earth's orbit, combined with the shorter changes produced by the precession of the equinoxes. The latter cause is due to the fact that the earth is not an exact sphere but slightly protuberant at the equator, and that the attraction of the sun on this protuberant matter prevents the axis round which the earth rotates from remaining exactly parallel with itself, and makes it move slowly round its mean position just as we see in the case of a schoolboy's top, which reels round an imaginary upright axis while spinning rapidly. This revolution in the case of the earth completes its circle in about 21,000 years, so that if summer, when the pole is turned towards the sun, occurred in the Northern hemisphere when the earth was in perihelion, or nearest the sun, and consequently winter when it was in aphelion, or furthest away from the sun, after 10,500 years the position would be exactly reversed, and winter would occur in perihelion and summer in aphelion ; the Southern hemisphere then enjoying the same conditions as those of the Northern one 10,500 years earlier. And in another 10,500 years things would come back to their original position.

Now if the earth's orbit were an exact circle this would make no difference, all the four seasons would be of the same duration and would receive the same solar heat in both hemispheres, and if the orbit were nearly circular, so that the difference between the perihelion and aphelion distances was small, the effect would be small also. But if the orbit flattened out or became more eccentric, the effect would be increased. The time of traversing the aphelion portion of the annual orbit would become longer and that of traversing the perihelion portion shorter, as the orbit departed from the form of a circle and became more elliptic. Whenever, therefore, the North Pole was turned away from the sun in aphelion, the winters would be longer than the summers in the Northern hemisphere, and conversely, the summers would be longer than the winters when, after an interval of 10,500 years, precession brought about the opposite condition of things, in which winter occurred in perihelion.

At present the earth's orbit is nearly circular, and the Northern hemisphere is nearest the sun in winter and farthest from it in summer, but the difference is only about 3,000,000 miles, or a small fraction of the total mean distance of 93,000,000 miles, which makes the winter half of the year shorter than the summer half by nearly eight days.

But mathematical calculations show that under the complicated attractions of the sun, moon, and larger planets, the eccentricity of the earth's orbit slowly changes at long and irregular intervals, but always within fixed limits, increasing up to a certain point and then diminishing till it approaches the circular form, when it again increases. The *maximum* limit of eccentricity makes the difference between the greatest and least distances of the earth from the sun range between 12,000,000 and 14,000,000 miles, which is four or five times as great as at present; and with this eccentricity, and winter in aphelion in the Northern hemisphere, the winter half of the year in Northern latitudes would be twenty-six days longer than the summer half, instead of eight days shorter as at present. In this state of things the quantity of heat received daily from the sun in winter would be such as to lower the temperature of the whole Northern hemisphere by 35° Fahrenheit, and reduce the average January temperature of England from 39 to 4°, while the mean summer temperature would be about 60° higher than at present. But this summer heat, derived from solar radiation, would not counteract the cold of winter, for all moisture during winter being accumulated in ice and snow, most of the solar heat of summer would be expended in supplying latent heat to melt a portion of this frozen accumulation, and dense fogs would intercept a large amount of the solar radiation.

After 10,500 years this state of things would be entirely reversed, and with twenty-six days more of summer, and the earth 12,000,000 miles nearer the sun in winter, the Northern hemisphere would enjoy something like perpetual spring. There can be no doubt that these are real causes, and the only difficulty is to account for their not having been more invariable in their operation and given us a constant succession of Glacial periods since the commencement of geological time, whenever the eccentricity became great, which occurs at irregular periods, but practically about three times in every 3,000,000 years. The answer is that the effects would only occur when the other conditions were present, viz., high land, moist winds, and an absence of oceanic currents of warm water like the Gulf Stream. The latter is one of the main causes which affect temperature. The difference of temperature between the equatorial and polar regions causes a constant overflow of heated air from south to north, which is replaced by an indraught of colder air from north to south, which, owing to the greater velocity of the earth's rotation towards the equator, takes the form of trade-winds blowing constantly from a more or less easterly direction. These winds, sweeping over the Atlantic Ocean, raise its level at its western barrier, and the accumulation deflected by America flows off in a current which extends to the western shores of Europe and carries mild winters into the extreme North. In the Orkney and Shetland Islands, which are nearly in the same latitude as Cape Farewell in Greenland, there is so little ice that skating is a rare accomplishment, and curling, the roaring game which is so popular some degrees further south, is quite unknown. If the Gulf Stream were diverted, and the highlands of Scotland upheaved to the height of the Alps of New Zealand, the whole country would again be buried under glaciers pushing out into the Atlantic and German Ocean.

These considerations may show why every period of

great eccentricity was not necessarily a Glacial period, though under certain conditions it must inevitably have been so, and geologists are generally agreed that the last period of the sort must have been one of the main causes of the great refrigeration which set in over the whole Northern hemisphere towards the close of the Pliocene period, and continued until recent times. But in this case we can fix the date with great accuracy, for calculation shows that the last period of great eccentricity began 240,000 years ago, and lasted 160,000 years. For the last 50,000 years the departure of the earth's orbit from the circular form has been exceptionally small. We may suppose the Glacial period, therefore, to have commenced 240,000 years ago, come to its height 160,000 years ago, and finally passed away 80,000 years before the present time.

These dates receive much confirmation from conclusions drawn from a totally different class of facts. A bed of existing marine shells of Arctic type, apparently belonging to one of the latest phases of the Glacial period, has been found on the top of a hill in North Wales which is now 1,100 feet above the sea-level, and the same marine drift seems to extend to a height of upwards of 2,000 feet. There must, therefore, have been a depression of the land sufficient to carry it many fathoms below the sea, and a subsequent elevation sufficient to carry the sea bottom up to a height of certainly 1,100 and probably over 2,000 feet. In all probability, these movements were very slow and gradual, like those now going on in Greenland and Scandinavia, for there are no signs of earthquakes or volcanic eruptions in the district;

and it is probable that pauses occurred in the movements, and a long pause when subsidence had ceased before elevation began. Without taking these pauses into account, and assuming the elevation only just completed, and that Sir C. Lyell's average of two and a half feet a century is a fair rate for these slow movements, it would have required 50,000 years of continued elevation to bring these shells, and 80,000 years to bring the marine drifts, up to their present height above the sea; and a similar period previously must be allowed for their submergence. We may fairly conclude, therefore, that upwards of 100,000 years have elapsed since these shells lived and died at the bottom of the sea towards the close of the Glacial period, which corresponds very well with the date assigned by astronomical calculations.

Again, another attempt to fix a date for the close of the Glacial period has been made by Monsieur Forel, a Swiss geologist, from actual measurements of the quantity of suspended matter poured into the Lake of Geneva by the Rhone, and the area of the lake which has been silted up since it was filled by ice. It is evident that this silting up at the head of the lake could only begin when the great Rhone glacier, which once extended to the Jura Mountains, had shrunk back into its valley far enough to pour its river into the lake. M. Forel's calculations give 100,000 years as the probable time required for the river to silt up so much of the lake as is now converted into dry land. The data are somewhat vague, as on the one hand the rate of deposition may have been greater when a large mass of ice and snow was being melted, while on the other hand it may have been less, while the glacier still occupied the valley almost to the head of the lake and the Rhone had only a course of a few miles. All that can be said, therefore, is that it gives an approximate date for the close of the Glacial period which, like that derived from rates of depression and elevation, corresponds wonderfully well with the date required by Croll's theory.

Now, whether the date be a little more or a little less, it is clear that man existed on earth throughout a great part, if not the whole, of the Glacial period. He had existed a long while in conjunction with a fauna of more Southern and African aspect, before the reindeer migrated in vast herds into Southern France. His remains are found in caves and river drifts associated with those of hippopotamus, an animal which could by no possibility have lived in rivers which for half the year were bound hard in ice. Such remains must therefore of necessity date either from a period before the great cold had set in, or from some inter-glacial period prior to the great cold which drove the reindeer, musk ox, glutton, and Arctic hare as far south as the slopes of the Pyrenees.

In England we can trace distinctly at least four successions of boulder clays, that is of the ground moraines of land ice, separated by deposits of drifts, sands, and brick-earths, formed while the glaciers were retreating and melting; and a number of the Palæolithic implements have been found in what was undoubtedly part of the period of the second or great chalky boulder clay, which overspreads the southern and eastern counties of England up to the Thames Valley. The discovery of Palæolithic remains in the deposit of St. Prest, near Chartres, makes it almost certain that some at least of the ruder instruments must date back to the very beginning of the Glacial period, and all the evidence points to the conclusion that man was living during the many alternations of climate of that period, and whenever the glaciers retreated, followed them up closely.

Thus far we have been going on certain and ascertained facts, confirmed by such numerous and well-authenticated proofs that doubt is impossible. But we get on less certain ground when we try to trace back human origin to more remote periods. As regards this question, we must begin by describing shortly the geological periods during which the existence of man may have been possible. It is useless to go back beyond the Chalk, which was deposited in a deep ocean and forms a great break between the modern and the Secondary period, in which latter reptiles predominated, and mammalia are only known by a few remains of small insectivorous and marsupial animals.

The inauguration of the present state of things commences with the Tertiary period. This has been divided into three stages: the Eocene, in which the first dawn appears of animal life similar in type to that now existing; the Miocene, in which there is a still greater approximation to existing forms of life; and the Pliocene, in which existing types and species become preponderant. Then comes the Pleistocene or Quaternary, including the great Glacial period, during which the whole marine and nearly the whole terrestrial fauna are of existing or recently extinct species, though very different in their geographical distribution from that of the present day. And finally we arrive at the recent period, when the present climate and the present configuration of lands, seas, and rivers, prevail with very slight modifications, and no changes have taken place either in the specific character or geographical distribution of life, except such as can be clearly traced to existing causes such as the agency of man.

This is the geological frame-work into which we have to fit the history of man's appearance upon earth. We have traced him through the recent and Quaternary, can we trace him further into the Tertiary? Speaking generally we may say that the Eocene period was that in which Europe began to assume something like its present configuration, and in which mammalian life, of the higher or placental type, began to supplant the lower forms of marsupial life which had preceded them. But these higher types were for the most part of a more primitive or generalised character than the more specialised types of later periods, and the highest order, that of the *primates*, which includes man, ape, and lemur, was, as far as is yet known, represented only by two or three extinct lemurian forms.

The plan on which Nature has worked in the evolution of life seems always to have been this: she begins by laying down a sort of ground plan, or generalised sketch of a particular form of life, say, first of vertebrata, then of fish, then of reptiles, and finally of mammalian life. This sketch resembles the simple theme of a few notes on which a musician proceeds to work out a series of variations, each surpassing the other in complication and specialised development in some particular direction. Now, in the Eocene period we are in the stage of the theme and first simple variations of the mammalian melody. It hardly seems likely, therefore, that a creature so highly specialised as man, even in his most rudimentary form, should have existed, and in the absence of any direct evidence to the contrary, it is safe to assume that his first appearance must have been of later date.

But when we come to the Miocene and Pliocene periods, the case is different. It is true that in the Miocene the specialisation of certain families, as for instance that of the horse, had not been carried out to the full extent, and that all the species of Miocene land-mammals and several of the genera are now extinct. But there were already true apes and baboons, and even two species of anthropoid ape, one of which, the Dryopithecus, whose fossil remains were found in the South of France, was as large as a man, and has been considered by some anatomists as in some respects superior to the chimpanzee or gorilla.

Now, wherever anthropoid apes lived it is clear that, whether as a question of anatomical structure or of climate and surroundings, man, or some creature which was the ancestor of man, might have lived also. Anatomically speaking, apes and monkeys are as much special variations of the mammalian type as man, whom they resemble bone for bone and muscle for muscle, and the physical animal man is simply an instance of the quadrumanous type specialised for erect posture and a larger brain. The larger brain, implying greater intelligence, must also have given him advantages in contending with outward circumstances, as for instance, by fire and clothing against cold, which might enable him to survive when other species succumbed and became extinct.

If he could survive, as we know he did, the adverse

conditions and extreme vicissitudes of the Glacial period, there is no reason why he might not have lived in the semi-tropical climate of the Miocene period, when a genial climate extended even to Greenland and Spitzbergen, and when ample forests supplied an abundance of game and edible fruits. The same reasons apply, with still greater force, to the Pliocene period, when existing types and species had become more common and when a mild climate still prevailed. The existence of Tertiary man must antecedently be pronounced highly probable; but probabilities are not proofs, and the fact of such existence must be determined by the evidence. All that can be said is that while there ought to be great caution in admitting as established a fact of such importance, there ought to be no determined predisposition to disbelieve it, like that which for so many years retarded the acceptance of the evidence for Palæolithic man. On the contrary, the fact that man existed in such numbers and under such conditions as have been described in the Quaternary period, establishes a strong presumption that his first appearance must date from a much earlier period.

Let us see how the evidence stands. Undoubted stone implements, and bones bearing traces of cuttings by flint knives, have been found in strata at St. Prest, near Chartres, which were always considered to be Pliocene. Since the discovery, however, some geologists have contended that these strata are not Pliocene, but of the earliest Quaternary or perhaps a transition period between Pliocene and Quaternary. This evidence cannot, therefore, be accepted as conclusive for anything more than proof that man's existence extends at any rate over the whole Quaternary period, comprising the vast glacial and inter-glacial ages which have effected such changes in the earth's surface.

The next piece of evidence is from Italy, where bones of the Balænotus, a sort of Pliocene whale, have been discovered in strata undoubtedly Pliocene, which bear marks of incisions which to all appearance must have been made by flint knives employed in hacking off the flesh. Doubts were thrown at first on this, as it was thought that possibly fish, or some gnawing animal

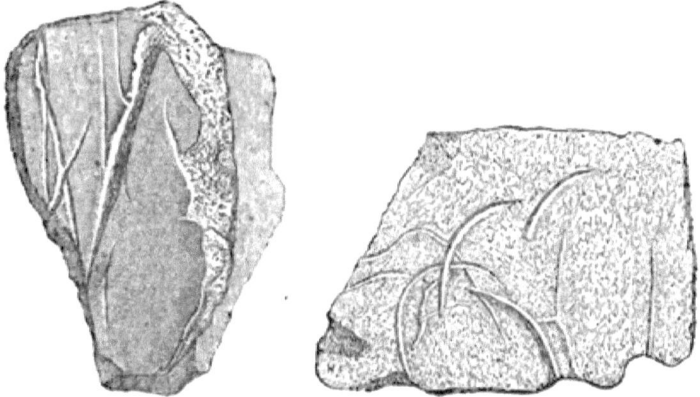

INCISED BONES OF BALÆNOTUS. Pliocene. From Monte Aperto, Italy.
Figured by Quatrefages, "Hommes Fossiles et Hommes Sauvages," p. 93.

like the beaver, might have cut the grooves with their teeth. But later specimens have been found on which the cuts have a regular curvature which could not have been made by any teeth, and present precisely the same appearance as the cuts which are so commonly found on the bones of reindeer and other animals in hundreds of Palæolithic caves.

M. Quatrefages, who is a very eminent and at the same time very cautious authority, says, in his last work on the subject published in 1884, "Hommes Fossiles et Hommes Sauvages," that "the most incredulous must be convinced. The hand of man armed with a cutting instrument could alone have left marks of this sort on a plain surface. It is evident that some horde of savages of these remote times has found the carcase of this great cetacean stranded on the shore, and cut the flesh off with stone knives just as the savages of Australia do at the present day." In fact incredulity only exists because this is as yet a solitary instance of Pliocene man, and scientific men, feeling that if true, further evidence must soon be found, very properly endeavour to keep their judgment in suspense.

If these bones of the Balænotus really bear marks of human tools, the spectacle which might have been witnessed on the shore of the Pliocene sea perhaps 500,000 years ago, must have closely resembled that given by Sir John Lubbock from a description by Captain Grey of a recent whale feast in Australia. " When a whale is washed on shore it is a real godsend to them. Fires are immediately lit, to give notice of the joyful event. Then they rub themselves all over with blubber, and anoint their favourite wives in the same way; after which they cut down through the blubber to the beef, which they sometimes eat raw and sometimes broil on pointed sticks. As other natives arrive they 'fairly eat their way into the whale, and you see them climbing in and about the stinking carcase, choosing titbits.' For days 'they remain by the carcase, rubbed from head to foot with stinking blubber, gorged to repletion with putrid meat —out of temper from indigestion, and therefore engaged in constant frays—suffering from a cutaneous disorder by high feeding—and altogether a disgusting spectacle. There is no sight in the world,' Captain Grey adds, 'more revolting than to see a young and gracefully-formed native girl stepping out of the carcase of a putrid whale.'"

The evidence for Miocene man is much of the same character: very strong and conclusive as far as it goes, but resting on too few instances to be universally accepted. In 1868 the Abbé Bourgeois laid before the Anthropological Congress at Paris certain flints which he had found *in situ* in undoubted Miocene strata at Thenay, in the Beauce, near Blois. They were received with general incredulity, and the traces of human design were denied. The Abbé, however, persisted, and having made fresh discoveries the subject was referred to the next meeting of the Congress at Brussels, who appointed a commission of fifteen of the most eminent European authorities in such matters to report upon it. Nine reported that some of the flints showed undoubted traces of human workmanship, five were of an opposite opinion, and one was neutral. Since then fresh objects have been found, and M. Quatrefages, who had formerly been doubtful, says in his recent work: "These new objects, and especially a scraper which is one of the most distinctly characterised of that class of implements, have removed my last doubts." And certainly, if the figures given at page 92 of his "Hommes Fossiles et Hommes Sauvages" correctly represent the original implements, and they really came from Miocene strata, doubt is no longer possible. The evidence of design in chipping into a determinate shape is quite as clear as in the similar

FLINT SCRAPER.
From Thenay. Miocene.
Figured by Quatrefages,
"Hommes Fossiles et
Hommes Sauvages,"
p. 92.

MIOCENE IMPLEMENTS FROM THENAY COMPARED WITH
UNDOUBTED PALÆOLITHIC IMPLEMENTS FROM QUATERNARY
CAVES AND DRIFTS.

MIOCENE.

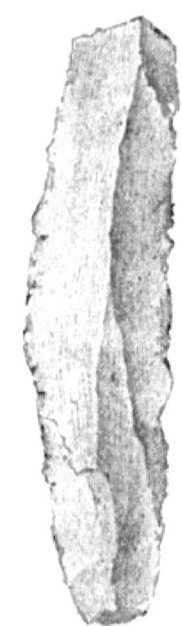

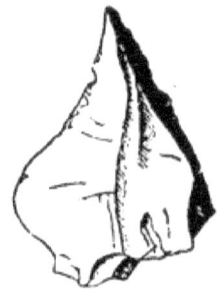

BORER, OR AWL.
Thenay. Miocene.
Congrès Préhistorique,
Bruxelles, 1872.

QUATERNARY. Chaleux,
Belgium. Reindeer Period.
Congrès Préhistorique,
Bruxelles, 1872.

SCRAPER, OR RUDE
KNIFE. Thenay. Miocene. Quatrefages,
p. 92.

SCRAPER. Thenay. Miocene.
Quatrefages, p. 92.

QUATERNARY.
From Le Moustier.

QUATERNARY. Mammoth Period.
River Drift, Mesvin, Belgium.
Congrès Préhistorique, Bruxelles, 1872.

class of implements from Kent's Cavern or the Cave of La Madeleine. They must either have been chipped by man, or as Mr. Boyd Dawkins supposes, by the Dryopithecus or some other anthropoid ape which had a dose of intelligence so much superior to the gorilla or chimpanzee as to be able to fabricate tools. But in this case the problem would be solved and the missing link discovered, for such an ape might well have been the ancestor of Palæolithic man.

The next instance is from the valley of the Tagus, where flint implements were alleged to have been discovered by an eminent Portuguese geologist, Señor Ribeiro, in Miocene strata. The subject was fully discussed on the spot, at a meeting of the Anthropological Congress at Lisbon in 1880. The general opinion seemed to be that some of the implements showed undoubted traces of human design, but some good authorities remained sceptical; and although there was no doubt that they were found in Miocene strata, it was thought possible that flints of Quaternary age might

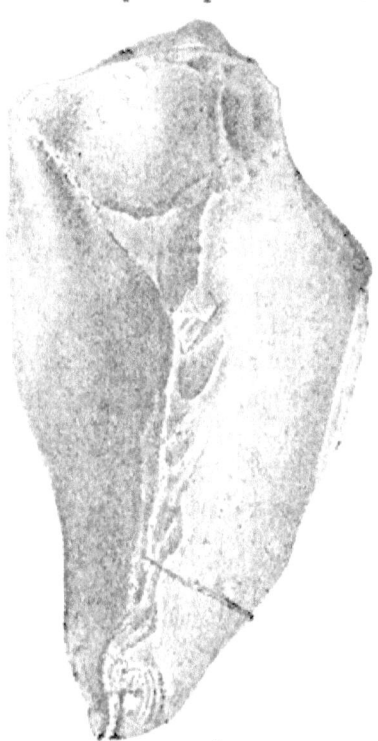

TERTIARY HÂCHE.
From Miocene Strata of Tagus Valley.
(Half the actual size.)
Quatrefages, "Hommes Fossiles et Hommes Sauvages."

have fallen into fissures, or been mixed up with Miocene sands by floods at some very remote period, and thus become encrusted in a Miocene matrix.

The verdict here, therefore, must be " Probable, but not proven." The same will apply to the alleged discovery of a human skull in California, buried under six distinct layers of hardened volcanic ashes, and certainly of Pliocene date, if not earlier. Whitney, the Director of the Geological Survey of the United States, and other American geologists, believe this skull to be Pliocene, but doubts have been thrown on its authenticity, and European geologists do not generally accept it.

A human bone is described by Lyell, which was found near Vicksburg in a side valley of the Mississippi, associated with bones of the extinct Mastodon and Megalonyx. But, although undoubtedly of great antiquity, there is no proof that it does not belong to the Quaternary period, especially as the mastodon seems to have lived until comparatively recent times in America, its remains being often found in recent bogs and peat mosses.

The same remark will apply to the skull which was found, in digging a well at New Orleans, under six distinct layers of cypress forests such as are now growing on the surface, showing as many periods of successive subsidences, subsequent elevations, and stationary periods long enough to allow of a forest growth of many generations of large trees. Here again the antiquity must be very great, but we have no reason to carry it back into Tertiary periods, or beyond the recent period when the Mississippi began to flow in its present course and form its present delta

Human remains have also been discovered in caves in Brazil associated with bones of extinct animals, but we have no clear information as to the time when these animals became extinct, or as to the exact order of superposition in which the human skulls and implements were found, and the occurrence of a polished stone celt in the same cave throws still more doubt on their extreme antiquity.

The existence of Tertiary man must for the present be considered as resting on three instances:

1. The undoubted flint implements and cut bones (including those of the *Elephas meridionalis*, a Pliocene and Miocene species) of St. Prest.
2. The cut bones of the Balænotus from the Pliocene strata of Monte Aperto in Italy, the cuts on which appear to have been undoubtedly made by the hand of man armed with a sharp cutting stone implement.
3. The flints from the Miocene strata of Thenay, some of which show unmistakable signs of having been split by fire and chipped into shape by design.

On the other hand the evidence is entirely negative, that a large number of fossil animal remains have been found in various parts of the world, specially in the Pliocene of the Cromer forest bed, and the Miocene of the Sewalik hills, Pikermi and Nebraska, without finding any trace of man. This is true, and is sufficient to make us require great caution in admitting as fully established a fact of so much importance, which would carry back the antiquity of man from one or two hundred thousand years to at least a million. But the example of Quaternary man shows the danger of trusting too exclusively to negative evidence. Thirty years ago the negative evidence against his existence was considered conclusive. Now his remains have been found over the whole world and in thousands of instances.

It must be remembered, also, that remains of Tertiary man are not likely to be abundant. If man was then living, it was probably in fewer numbers and in more limited areas. The pressure of population had not yet driven wandering hordes to follow sea-coasts and cross rivers and mountains in pursuit of food. Probably at this early period man lived more on fruits, and therefore required fewer implements, and his intelligence was less, so that he had less power of fashioning them. For the purposes for which his Palæolithic descendants chipped stones into shape, he may have used natural stones which would often answer the purpose, but which, when thrown away, would leave nothing by which they could be recognised.

If the forests now inhabited by the gorilla and chimpanzee were submerged and again elevated, no trace would be found of the existence of animals which had built rude nests, used broken branches of trees as clubs, and cracked cocoa-nuts with hammer stones.

But above all, the surface of these older strata has been so much denuded, that the situations in which alone we might expect to find remains of man have almost entirely disappeared. Ninety-nine hundredths of our Quaternary implements come from river drifts or caves. Where are the Pliocene or Miocene rivers or caves? They have disappeared amidst the revolutions of the earth's surface and the constant denudation which wastes continents away. The negative evidence would be strong if we could point to caves filled with bone-breccias of a Pliocene or Miocene fauna, in which no trace was found of human remains. But it is weak as against even a single well-ascertained instance, if it merely amounts to such remains not being frequently found where we could hardly expect to find them. And it is weak against the strong presumption that when Quaternary man is found in such numbers and under such conditions, spread over wide areas in inhospitable climates, he must have had his first origin in earlier times. It is, therefore, in the highest degree probable that this origin must have been in Tertiary times, when we know as a certain fact that large anthropoid apes were already in existence.

If this were so, what would it teach us as to the date of man's appearance?

Reckoning by the thickness of the different stratified deposits which make up the earth's crust, and assuming the average rate of their deposition, or what is the same thing, the average rate of waste of land surface to have been the same throughout, the whole Tertiary period carries us back barely one-twentieth part of the way towards the first beginnings of fossil-bearing strata. That is, if 100,000,000 years have elapsed since the earth became sufficiently solidified to support vegetable and animal life, the Tertiary period may have lasted for 5,000,000 years; or for 10,000,000 years, if the life-sustaining order of things has lasted, as Lyell supposes, for at least 200,000,000 years. Even if we take the shorter period, the time is ample for the enormous changes which have taken place since the commencement of the Eocene period. The average rate of denudation over the globe has been taken at about one foot in 3,000 years, from actual calculations of the average amount of solid matter carried down by the Mississippi and other great rivers. Now at this rate it would take only 2,000,000 years to wear the whole of Europe down to the sea-level, and, in the absence of any compensating movements of elevation, the whole of North America would be washed away and deposited in strata at the bottom of the Atlantic and Pacific Oceans in less than 3,000,000 years.

If, therefore, the origin of man could be traced down to the middle Miocene, or even to the date of the great anthropoid Dryopithecus of Southern France, we should have to assume a period for his existence of probably between one and two millions of years, a mere fraction of the time since the earth became the abode of life and existing causes operated to bring about geological formations.

As regards the habits and manners of Quaternary man we know very little that is positive, and can only gather some vague indications from the relics of caves and river drifts. These, however, are sufficient to establish with certainty that the law of his existence has been one of continued progress. The older the remains, the ruder are the implements and the fewer the traces of anything approaching to civilisation. In the Neolithic period man is comparatively civilised. He has domestic animals and cultivated plants; he has clothing and ornaments, well-fashioned tools and pottery, and permanent dwellings. He lives in societies, builds villages, buries his dead, and shows his faith in a future life by placing with them food and weapons. As we ascend the stream of time these indications of an incipient civilisation disappear. The first vestige of the domestic animals is found in the dog which gnawed the bones of the Danish kitchen-middens, and of the earliest Swiss lake-dwellings. When fairly in Palæolithic times even the dog disappears, and man has to trust to his own unaided efforts in hunting wild animals for food.

Weapons and implements become more and more rude until, in the oldest deposits, we find nothing but roughly-chipped hatchets, arrow-heads, flakes, and scrapers. Implements of bone, such as barbed harpoons, borers, and needles, which are abundant in the middle Palæolithic or reindeer period, become ruder and disappear. Pottery, which is extremely abundant in the Neolithic period, either disappears altogether or becomes so scarce that it is a moot question whether a few of the rudest fragments found in caves are really Palæolithic. If so, they clearly date from the later Palæolithic, and pottery was unknown in the earlier Palæolithic times.

Judging from the portraits engraved on bone during the reindeer period, Palæolithic man pursued the chase in a state of nature, though from the presence of bone needles it is probable that the skins of animals may have been occasionally sewed together by split sinews to provide clothing. There can be no doubt that his habitual dwelling was in caves or rock-shelters. Here was his home, here he took his meals and allowed the remains of his food to accumulate. His staple diet consisted of the contemporary wild animals, the mammoth, the rhinoceros, the cave bear, the horse, the aurochs, and the reindeer. Even the great cave lion was occasionally killed and eaten, and the fox and other smaller animals were not despised; while among tribes skilled in the use of the bow and arrow, birds were a common article of food, and fish were harpooned by those who lived near rivers. Wild fruit and roots were also doubtless consumed, and from the formation of his teeth and intestines it is probable that if we could trace the diet of the earliest races of men we should find them to have been frugivorous, like their congeners the anthropoid apes.

The abundance of wild animals and the long period for which hunting savages inhabited the same spots may be inferred from the fact that at one station alone, that of Solutré in Burgundy, it is computed that the remains of no less than 40,000 horses have been found. All the long bones of the larger animals have been split to extract the marrow, which seems, as with the modern Eskimos and other savages, to have been a great delicacy, and also used for softening skins for the purpose of clothing.

Among the split bones a sufficient number of human bones have been found to make it certain that Palæolithic man was, occasionally at least, a cannibal; and in several caves, notably that of Chaleux, in Belgium, these bones, including those of women and children, have been found, charred by fire, and in such numbers as to indicate that they had been the scene of cannibal feasts. It is a remarkable fact that cannibalism seems to have become more frequent as man advanced in civilisation, and that while its traces are frequent in Neolithic times, they become very scarce or altogether disappear in the age of the mammoth and the reindeer.

As regards religious ideas they can only be inferred from the relics buried with the dead, and these are scarce and uncertain for the earlier periods. The caves in which Palæolithic man lived on the flesh of the Quaternary animals, have been so often used as burying-places in long-subsequent ages, that it is extremely difficult to ascertain whether the skeletons found in them are those of the original inhabitants. Thus the famous cave of Aurignac, in which Lartet thought he had discovered the tomb of men at whose funeral feast mammoths and rhinoceroses were consumed, is now generally considered to be a Neolithic burying-place superimposed on an abandoned Palæolithic habitation.

There are not more than five or six well authenticated instances in which entire Palæolithic skeletons have been found under circumstances in which there is a fair presumption that they may have been interred after death, and these afford no clear proof of articles intended for use in a future life having been deposited with them. All we can say, therefore, is that from the commencement of the Neolithic period downwards, there is abundant proof that man had ideas of a future state of existence very similar to those of most of the savage tribes of the present day; such proof is wanting for the immensely longer Palæolithic period, and we are left to conjecture. The only arts which can with certainty be assigned to our earliest known ancestors are those of fire and of fashioning rude implements from stone by chipping. Everything beyond this is the product of gradual evolution.

CHAPTER VI.

MAN'S PLACE IN NATURE.

Origin of Man from an Egg—Like other Mammals—Development of the Embryo—Backbone—Eye and other Organs of Sense—Fish, Reptile, and Mammalian Stages—Comparison with Apes and Monkeys—Germs of Human Faculties in Animals—The Dog—Insects—Helplessness of Human Infant—Instinct—Heredity and Evolution—The Missing Link—Races of Men—Leading Types and Varieties—Common Origin Distant—Language—How Formed—Grammar—Chinese, Aryan, Semitic, etc.—Conclusions from Language—Evolution and Antiquity—Religions of Savage Races—Ghosts and Spirits—Anthropomorphic Deities—Traces in Neolithic and Palæolithic Times—Development by Evolution—Primitive Arts—Tools and Weapons—Fire—Flint Implements—Progress from Palæolithic to Neolithic Times—Domestic Animals—Clothing—Ornaments—Conclusion, Man a Product of Evolution

ALTHOUGH the establishment of the great antiquity of the human race has attracted more immediate attention, being a fact at once intelligible to the general public, the researches of anatomists and physiologists, aided by the microscope, have brought to light results quite as remarkable as regards the individual man and his place in Nature. Until recently it was taken for granted that man was a special miraculous creation, altogether superior to and distinct from the rest of the animal world. This assumption, gratifying alike to our vanity and our laziness in the laborious search for truth, has been to a great extent disproved and replaced by the Law of Evolution.

The most striking proof of this is found when we trace scientifically the growth of each individual man from his first origin to his final development. Man, like all other animals, is born of an egg. The primitive egg, or ovum, which was the first germ of our existence, is a small cell about the one-hundredth of an inch in diameter, consisting of a mass of semi-fluid protoplasm enclosed in a membrane, and containing a small speck or nucleus of more condensed protoplasm. This nucleated cell is itself the first form into which a mass of simple jelly-like protoplasm is differentiated in the course of its evolution from its original uniform composition. The nucleated cell is the starting-point of all higher life, and by splitting up and multiplying repetitions of itself in geometrical progression, provides the cell material out of which all the complicated structures of living things are built up. In sexual generation, which prevails in all the higher forms of life, this process requires, in order to start it, the co-operation of two such cells or germs of life, one male, the other female.

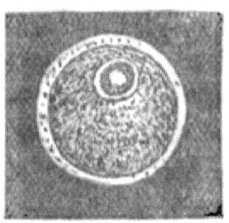
HUMAN EGG.
Magnified 100 times

The first remarkable fact is that the human egg is, at its commencement, undistinguishable from that of any other mammal, and remains so for a long period of its growth, going through its earlier stages of development in precisely the same way. At first the egg behaves exactly as any other single-celled organism, as for instance that of the amœba, which is considered the simplest form of organised life. It contracts in the middle and divides into two cells, each with its nucleus and each an exact counterpart of the original cell. These two subdivide into four, the four into eight, and so on, until at last a cluster of cells is formed which is called a *morula* from its resemblance to the fruit of the mulberry-tree. Development goes on, and the globular lump of cells changes into a globular bladder whose outside skin is built up of flattened cells. Then condensation takes place, from

MAMMALIAN EGG.
First Stage.   Second Stage.   Third Stage.

the more rapid growth of cells at particular points, and the foundation is laid of the actual body of the germ or embryo, the other cells of the germ-bladder serving only for its nutrition. Up to this point the germs not only of all mammals including man, but of all vertebrate animals, birds, reptiles, and fishes, are scarcely distinguishable.

In the next stage the outer surface of the embryo develops three distinct layers, the outer one of which, or epidermis, becomes the outer skin; the inner one, or epithelium, the mucous membrane or lining of all the intestinal organs; and the intermediate layer the raw material of muscles, bones, and blood-vessels.

The embryo is now contracted in the middle and assumes the form of a violin-shaped disc, and a slight longitudinal furrow appears, dividing it into two equal right and left parts, which is gradually converted into a tube containing the spinal marrow, to protect which a chain of bones or vertebræ is developed, forming the back-bone.

And now comes what is the most marvellous part of the process, viz., the development of the brain, eye, ear, and other organs of sense, from these simple elements. The brain begins as a swelling of the foremost end of the cylindrical marrow-tube. This divides itself into five bladders, lying one behind the other, from which the whole complicated structure of the brain and skull is subsequently developed.

The eye, ear, and other sense-organs, begin in the same way. A slight depression in the outer skin extends until the edges close and form a hollow space in which the eye is formed. At first it is a mere black pigment mark on the interior surface of the enclosed space, which develops into the retina, with a wonderful apparatus of optic nerves for conveying impressions photographed on it to the brain. The enclosed space itself is filled with a fluid, or vitreous humour, from which a lens is condensed for collecting the rays of light and concentrating them on the retina, and by degrees all the beautiful and complicated organs are evolved for perfecting the work of the eye and protecting it from injury. But this fact must be kept clearly in view: the process is identically the same as that by which the eyes of other animals are formed, and its various stages represent those by which the organs of vision have gradually risen to the development of a complete eye, in advancing from the lowest to the higher forms of life. Thus in the lowest, or Protista, the eye remains a simple pigment spot, which probably perceives light by being more sensitive to variations of temperature than the surrounding white cells. The next higher family develop a lens, and so on in ascending order, different families developing different contrivances for attaining the same object, but all starting from the same origin, development of the cells of the epidermis, and leading up to the same result, organs of vision adapted for the ordinary conditions of life of the creature which uses them. I say the *ordinary* conditions, for there are curious instances of the eye persisting, dwindling from disuse, and finally disappearing, in animals which live underground like the mole, or in subterranean waters like some fish in the Mammoth Cave of Kentucky and underground lakes of Carinthia, where the stimulus of light is no longer felt for many generations.

The history of the ear and other organs of sense is the same as that of the eye. They are all developments of the cell system of the outer skin, and all pass through stages of development identical with those at which it has been arrested in the progression from lower to higher forms of life. The same principles apply to the development of the inner organs, such as the heart, lungs, liver, etc., a striking illustration of which is found in the fact that the gill arches, or bones which support the gills by which fishes breathe, exist originally in man and all other vertebrate animals above the ranks of fish, but, in the development of the embryo, they are superseded by the air-breathing apparatus of lungs, and converted to other purposes in the formation of the jaws and organ of hearing. In fact, we may say that every human being passes through the stage of fish and reptile before arriving at that of mammal, and finally of man.

If we take him up at the more advanced stage, where the embryo has already passed the reptilian form, we find that for a considerable time the line of development remains the same as that of other mammalia.

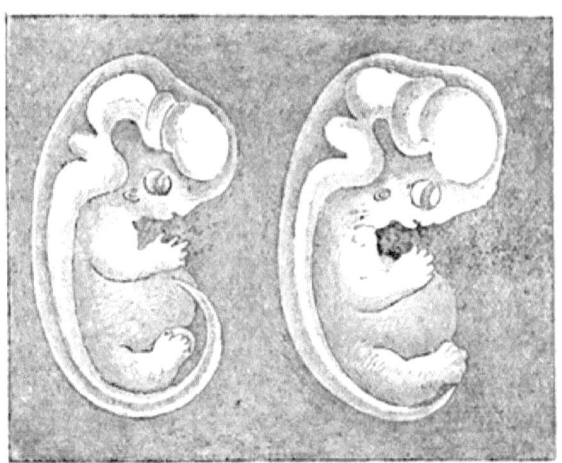

Dog (six weeks).   Man (eight weeks).
From Haeckel's "Schöpfungsgeschichte."

The rudimentary limbs are exactly similar, the five fingers and toes develop in the same way, and the resemblance after the first four weeks' growth between the embryo of a man and a dog is such that it is scarcely possible to distinguish them. Even at the age of eight weeks the embryo man is an animal with a tail, hardly to be distinguished from an embryo puppy.

As evolution proceeds the embryo emerges from the general mammalian type into the special order of *Primates* to which man belongs. This order, beginning with the lemur, rises through the monkey, the baboon, and tailed ape, up to the anthropoid apes, the chimpanzee, gorilla, and orang, which approach nearest to the human type. The succession is gradual from the lower to the higher forms up to the anthropoid apes, but a considerable gap occurs between these and man. It is true that in his physical structure man resembles these apes closely, every bone and muscle of the one having its counterpart in those of the other. But even at its birth the human infant is already specialised by considerable differences. The brain is larger, its convolutions more complex, the spine has a double curvature, adapting it for an erect posture, and the legs, with a corresponding object, are longer and stronger, while the arms are shorter and less adapted for climbing. The thumb also is longer, making the hand a better instrument for all purposes, except that of clasping the branches of trees, for which the long, slender fingers of the ape are more available. The great toe also is less flexible and the foot more adapted for giving the body a firm support and less for being used as a hand.

As growth proceeds after birth these differences become more and more accentuated. The infant chimpanzee is not so very unlike the infant negro, but after a certain age the sutures of the skull close in the former, making the skull a solid box, which prevents further expansion of the brain, and the growth of the bone is directed towards the lower part of the face, giving the animal a projecting muzzle, massive jaws, and a generally bestial appearance, while at the same time its intelligence is arrested and its ferocious instincts become more prominent. Still these higher apes remain creatures of very considerable intelligence and warm affections, as may be seen in the behaviour of those which have been caught young and brought up under the influence of kind treatment. There is a chimpanzee now in the Zoological Gardens at Regent's Park, which can do all but speak, which understands almost every word the keeper says to it, and when told to sing will purse out its lips and make an attempt to utter connected notes. In the native state they form societies, obey a chief, and often show great sagacity in their manner of foraging for food and escaping from danger.

Even in lower grades of life than the anthropoid apes we can see plainly many of the germs of human faculties in an undeveloped state. Those who are fond of dogs, and have lived much with them and understood their ways, must have been struck by the many human-like qualities they possess, and especially by the very great resemblance between young dogs and young children. They both like and dislike very much the same people and the same mode of treatment. They like those who take notice of them, caress them, talk to them, and, above all, those whom they can approach with perfect confidence of receiving uniform kind treatment. They dislike those who have no sympathy with them, or whose treatment of them is either cold or capricious. Their great delight is to play with one another, and often to tease and make a pretence of quarrelling and fighting. They both have an instinct for mischief, and are constantly trying it on how far they can go without getting into serious difficulties.

Later in life, and in more serious matters, the dog has certainly the germs of intelligence, and does a number of things which require a certain exercise of reasoning power. He has a good memory, and imagination enough to be excited at the prospect of a walk where there is a chance of finding a rat or a rabbit, and to dream of chasing imaginary rabbits when he is lying curled up on the hearthrug. Every dog has an individual character of his own as clearly defined as that of an individual man, nor can the rudiments of consciousness be denied to the hound who, in a kennel of twenty others, knows perfectly well that he is Rover, and not Rattler or Ranger, and waits till his name is called to come forward for a biscuit. When he has got it, his sense of property makes him appropriate it as his own, and respect the biscuits appropriated to other dogs, at any rate to the extent of knowing perfectly well that he is doing wrong if he takes them by force or steals them.

In moral qualities the dog approaches even more closely to man. His fidelity, affection, and devotion even to death, are proverbial. He feels shame and remorse when he has departed from the canine sense of right and wrong or from the canine standard of honour, and is happy when he feels that he has done his duty. What is this but the working of an elementary conscience? Even in the higher sphere of religious feeling, the dog feels unbounded love and reverence for the master who is the highest being conceivable to him, or in other words, his God; and he shudders as that master does in the presence of anything weird and supernatural. Every good ghost story begins by describing how the dogs howled and shrank to their master's feet when the first shadow of supernatural presence was cast on the haunted castle.

Capacity for progressive improvement can hardly be denied to a race which has developed such qualities from ancestors who, like the wild and half-wild dogs of Asia and America, had not even learned to bark, and were as unlike the civilised and affectionate collie, as Palæolithic man to his modern successor. In fact, the progress of the dog seems only to be limited by the want of organs of speech, and of an instrument like the hand by which to place himself in closer relation with the outer world.

The same remarks apply to the elephant, whose great sagacity seems clearly attributable to the possession of such an instrument in the trunk, inferior no doubt to the hand, but still very superior to the paw of the dog or to the hoof-enclosed fore-foot of the horse. In all animals the greater or less perfection of the instruments by which they act upon and are acted upon by the outer world, seems to be the principal factor in determining the quality of the brain as an organ of intelligence.

In the insect world we find still more wonderful exemplifications of the resemblance between animal and human intelligence. Ants live in organised societies, build cities, store up food for winter, keep aphides as milk-cows, carry on slave-hunting raids, and push the division of labour to such an extent that some tribes are all workers, others all warriors and slave-owners. These actions are not all merely mechanical and instinctive, for ants can to a considerable extent adapt themselves to circumstances, and alter their habits and mode of life when it becomes necessary in the "struggle for existence." The same is true of bees, beetles, and other insects, but it is useless to dwell on these, for the organisation of the insect world is so different from that of the mammalian, to which man belongs, that no safe analogy can be drawn from one to the other. It is from the higher mammalian types that we can fairly draw the inference that, if like effects are produced by like causes, the more perfect intelligence, consciousness, and morality of man, must be the same in kind though higher in degree than the less perfect manifestations of the same qualities in animals of similar though less perfect physical organisation.

There is one respect in which the human infant differs greatly from the young of other animals, viz., in the long period for which it remains in a condition of utter helplessness. In many of the lower forms of life the young creature emerges into the world with many of its necessary faculties complete, and has to learn comparatively little from education. The chicken runs about and picks up food on the day it escapes from the egg, and the young flycatcher will peck at flies with fragments of the shell still adhering to it. As we rise in the scale of creation, these instinctive aptitudes become fewer, and more time is required before the young animal can shift for itself; and at length, in the human infant, we arrive at a stage where for the first year or two it can do little to preserve its existence except to breathe and suck.

The reason of this is doubtless to be found in the higher development to which it is destined to attain. The faculties of every animal depend on two causes— first, heredity, or those which have been evolved from the type, and become fixed by succession through a long series of ancestors; secondly, adaptation, or those which are acquired by education, including in the term everything that is requisite to place the animal in harmony with its surrounding environment. The first are what are called instincts, which exist from the birth, and are preserved unconsciously and without an effort. The last involve an effort, and reference from the outer stations of the senses along the telegraph wires called nerves, to the central office of the brain, where the message is recorded and the reply considered and transmitted along another set of nerves to the muscles, where it translates itself into action. In either case the fundamental fact seems to resolve itself into a tendency of molecular motion to follow beaten rather than unknown paths. What the brain has once thought or perceived, it will think or perceive more readily a second time, and in like manner, a message which has once been transmitted and read off along a nerve, from muscle to brain or from brain to muscle, will be transmitted and read off more readily by practice, until at length it ceases to require conscious effort and becomes instinctive. We may see an illustration of this in the facility with which a piano player, who began by learning the notes with difficulty, acquires such aptitude that the execution of rapid passages becomes mechanical, and can be carried on without a mistake, even when the performer is thinking of something else or talking to a bystander.

The outer world with which every animal has to deal from its birth upwards, may be compared to a dense forest or jungle through which it has to find its way. A certain number of paths have been cut by its ancestors, and it finds them ready made by heredity; others it constructs for itself by repeated efforts until they become as broad and easy as those which it inherited; and finally, if the forest is thick and its area extensive it can only be explored by leaving the beaten paths of inherited or acquired instinct, and groping the way painfully by conscious effort and attention.

We can now see why the lower the animal, or in other words the less extensive the forest, the whole vital energy may be concentrated on the few beaten paths opened by heredity, and a few necessary actions may be performed from the first, instinctively and with great perfection, while in higher organisms the vital energy is employed in developing a great mass of future possibilities rather than a small number of inferior present realities. The baby cannot run about the room and feed itself like the chicken, because the baby has to grow into a man or woman, while the chicken has only to grow into a fowl which can do very little more in its adult than in its infant state.

In fact, when we come to analyse the sum of faculties of the adult man, we find that they are derived to a surprisingly small extent from heredity as compared with education. In saying this, however, it must be understood that the term "heredity" is limited to that direct heredity which transmits characters by instinctive necessity, and not to the far larger sphere of indirect heredity by which faculties, arts, modes of thought, and rules of conduct, are accumulated in civilised societies, and become the principal instrument of education in its larger sense. If it were possible to suppose a human infant born of civilised parents, left entirely to itself, what would it grow into? Perhaps it would learn to walk, though this is not quite certain, as the few wild children who have been discovered in forests, went very much on all fours, and if we can believe the accounts of wolf children in India, those educated among wolves adopt their gait and habits; certainly it would not learn to speak, in the sense of using any articulate language; its arts would not extend beyond recognising a few articles of food, and perhaps using stones to crack nuts, and constructing some rude shelter from branches of trees. It would know nothing of fire, and on the whole would not be so far advanced as its oldest Palæolithic ancestor.

As regards a moral sense, and all that we are accustomed to think the highest attributes of humanity, it is clear that his mind would be a blank. Even at a much more advanced stage, such ideas evidently come from education, and are not the results either of inherited instinct or of supernatural gift. An English child kidnapped at an early age by Apache Indians or head-hunting Dyaks, would, to a certainty, consider murder one of the fine arts, and the slaughter of an inoffensive stranger, especially if accomplished with a treachery that made the exploit one of little risk, an achievement of the highest manhood. If brought up among Mahometans he would consider polygamy, if among the Todas polyandry, as the natural and proper relation of the sexes. All that can be said is, that if recaptured and brought back to civilised society, he would perhaps be assisted by heredity in adopting its ideas more readily than would be the case if he had been born a savage.

It is clear, therefore, that the history of the individual man tells the same story of evolution from low beginnings as is told by that of the human race as traced from Palæolithic, through Neolithic, into modern times. His law is progress, worked out by conscious effort called forth by the environment of outward circumstances, and accelerated from time to time by the successful efforts of a few superior men, whose greater sum of energy or happier organisation for development, enables them to pioneer new paths through the vast unexplored forests of science, art, and morality.

The difficulty of accounting for the development of intellect and morality by evolution is not so great as that presented by the difference in physical structure between man and the highest animal. Given a being with man's brain and man's hand and erect stature, it is easy to see how intelligence must have been gradually evolved, and rules of conduct best adapted for his own good and that of the society in which he lived must have been formed and fixed by successive generations, according to the Darwinian laws of the "struggle for life" and the "survival of the fittest."

But it is not so easy to see how this difference of physical structure arose, and how a being came into existence which had such a brain and hand, and such undeveloped capabilities for an almost unlimited progress. The difficulty is this: the difference in structure between the lowest existing race of man and the highest existing ape is too great to admit of the possibility of one being the direct descendant of the other. The negro in some respects makes a slight approximation towards the Simian type. His skull is narrower, his brain less capacious, his muzzle more projecting, his arm longer than those of the average European man. Still he is essentially a man, and separated by a wide gulf from the chimpanzee or gorilla. Even the idiot or *crétin*, whose brain is no larger and intelligence no greater than that of the chimpanzee, is an arrested man and not an ape.

If, therefore, the Darwinian theory holds good in the case of man and ape, we must go back to some common ancestor from whom both may have originated by pursuing different lines of development. But to establish this as a *fact* and not a *theory* we require to find that ancestral form, or, at any rate, some intermediate forms tending towards it. We require to find fossil remains proving for the genus man what the Hipparion and Anchitherium have proved for the genus horse, that is, gradual progressive specialisation from a simple ancestral type to more complex existing forms. In other words, we require to discover the "missing link." Now it must be admitted that hitherto, not only have no such missing links been discovered, but the oldest known human skulls and skeletons, which date from the Glacial period, and are probably at least 100,000 years old, show no very decided approximation towards any such pre-human type. On the contrary, one of the oldest types, that of the men of the sepulchral cave of Cro-Magnon, is that of a fine race, tall in stature, large in brain, and on the whole superior to many of the existing races of mankind. The reply of course is that the time is insufficient, and if man and the ape had a common ancestor, that as a highly developed anthropoid ape certainly, and man probably, already existed in the Miocene period, such ancestor must be sought still further back, at a distance compared with which the whole Quaternary period sinks into insignificance. It is said also that the discovery of man's antiquity is of quite recent date, and that thirty years ago the same negative evidence was quoted as conclusive against his existence in times and places which now afford his remains by tens of thousands. All this is true, and it may well make us hesitate before we admit that man, whose structure is so analogous to that of the animal creation, whose embryonic growth is so strictly accordant with that of other mammals, and whose higher faculties of intelligence and morality are so clearly not miraculous instincts but the products of evolution and education, is alone an exception to the general law of the universe, and is the creature of a special creation.

This is the more difficult to believe, as the ape family which man so closely resembles in physical structure, contains numerous branches which graduate into one another, but the extremes of which differ more widely than man does from the highest of the ape series. If a special creation is required for man, must there not have been special creations for the chimpanzee, the gorilla, the orang, and for at least 100 different species of apes and monkeys which are all built on the same lines?

What are the facts really known to us as to man, his nature, and his origin?

Man is one of a species of which there are in round numbers some 1,200 millions of individuals living at the present time on the earth. Taking thirty years as the average duration of each generation there are thus over 3,000 millions who are born and die per century, and this has gone on more or less during the period embraced by history which extends for a great part of the Old World over thirty centuries, in the case of Assyria and China over forty or fifty, and in Egypt over seventy centuries. At the commencement of these historical periods population was dense, probably in Egypt and Western Asia denser than at present, and civilisation far advanced. The Pyramids, which are at the same time the oldest and the largest buildings in the world, prove this conclusively, both from the mechanical skill and astronomical science shown in their construction, and from the great accumulation of capital and highly artificial arrangements of society which could alone have rendered such works possible. The great mass of the population in these olden times lived in what is known as the Old World, and was accumulated mainly in the great valley systems of the Nile, and of the various rivers and irrigated plains of the southern half of the continent of Asia. Northern Asia and Europe were thinly inhabited by ruder tribes. Of America and the interior of Africa we know little until a much later date, but the population was in all probability sparse and savage, while in Australia, if it existed at all, it was still scantier and more savage; while in New Zealand and most of the Pacific Islands it has only been introduced by migration within comparatively recent times.

The next leading fact we have to observe is that the human race is not everywhere the same, but is divided into several well-marked varieties. The most obvious distinction is that of colour. In the Old World there are three distinct and clearly characterised groups —the white, the yellow, and the black. These are found mainly in three separate zoological provinces:

the white in the temperate and north-temperate zones of Europe and Western Asia, the yellow in those of Eastern Asia, and the black in the tropical zone, principally of Central Africa. Where they are pure and unmixed, these race-types differ from one another not in colour only but in many other important and permanent characters. The average size of the brain, the complexity of its convolutions, the shape of the skull, the bones of the face and jaws, the comparative length of the limbs, the structure of the hair and skin, the characteristic odour, the susceptibilities to various diseases, are all essentially different, so that no observant naturalist, or even observant child or dog, could ever mistake a Chinaman for a Negro, or a Negro for an Englishman.

Such a naturalist, seeing for the first time typical specimens of the three races, would pronounce them without hesitation to be distinct species, and would predict with much confidence that they would either not cross, or, if they did, would produce a hybrid progeny of inferior fertility.

But here he would be wrong, for, in fact, the most opposite races breed freely together, and produce a fertile progeny.

Moreover, when we extend our view beyond the clearly distinguished types of the white, yellow, and black, as seen in Caucasian, Mongoloid, and Negro races, we find these types breaking off into sub-types and shading off towards each other, while a large proportion of the human race consists of brown, red, olive, and copper-coloured people, who may either be original varieties, or descended from crosses between the primitive races. Small isolated groups also crop up, differing from the main races, of whom it is hard to say from whom they are descended or how they got there; as for instance the Hottentots, in South Africa, the pigmy black Negritos of the Andamans and other South Asiatic islands, the Papuans and Australians, the hairy Ainos of Japan, and some of the aboriginal races of India.

To a certain extent climate seems to have had an influence in creating or developing the main typical differences. Thus the main line of black races lies along the hot tropical belt of the earth from Old to New Guinea. But the rule is not universal, there is no similar type in tropical America, where a singular uniformity of type and colour prevails throughout the whole continent. Even in Africa we find the Negro type, while retaining its black colour, shading off towards higher types and losing its more animal-like characteristics. Again, while colour becomes generally lighter as we pass from tropical to south-temperate and from south to north-temperate regions, if we go still further north we find darker races, such as the Lapps and Esquimaux, and in one remarkable instance the colour within the temperate zone itself actually becomes darker with increase of latitude, and the aboriginal savage of Tasmania, in a climate like that of Devonshire, was blacker than many negroes.

Even within great and well-defined races themselves there are clearly marked varieties. Thus the white race consists of the two distinct types of the fair-whites and dark-whites, the former prevailing in Northern Europe and the latter in Southern Europe, Western Asia, and North Africa; the contrast between a fair Swede with flaxen hair and blue eyes, and a swarthy Spaniard with black hair and eyes, being almost as marked as between the latter and some of the higher black or brown races. Throughout a great part of Europe, including specially England, it is evident that the existing population is derived mainly from repeated crosses of these two races with one another and probably with earlier races.

In the existing state of things also it is evident that if the different races of mankind ever really did pass into one another under influences like those of climate, the time of their doing so is long past. A colony of English families transported to tropical Africa would to a certainty die out long before they had taken even the first step towards acquiring the black velvety skin, the woolly hair, the projecting muzzle, and the long narrow skull of the typical Negro, while a Negro colony transported to Scotland or Scandinavia would as certainly disappear from diseases of the chest and lungs, long before they began to vary towards the European type. The yellow race seems to be on the whole the best fitted to withstand climate and other external influences, and it certainly shows no signs anywhere of passing over either into the Caucasian or the Negro type.

On the whole, therefore, if the fact of fertile intercrossing is to be taken as proving the unity of the human race and their probable descent from a common ancestor, and we are to assume that all the great varieties which we find existing are the result of modifications gradually introduced by climate and surrounding circumstances, it is evident that the point of divergence must be put at an immense distance.

This is the more certain, as when we look back for

a period of more than 4,000 years, we find from the Egyptian monuments that some of the best-marked existing types have undergone no sensible change. The portraits of negroes and of Semitic dark-whites painted on the walls of temples and tombs of the 12th dynasty, about 2000 B.C., might be taken as characteristic portraits of the negro and Jew of the present day, and the modern Egyptian fellah reproduces with little or no change the features of the ancient Egyptians of the days of Rameses and Amenophis. It is evident, therefore, that where no great change has taken place from crossing of races, they will maintain their special characters unaltered for more than 100 generations. Indeed we might say for 200 generations, for the statues and wooden statuettes from the tombs of Sakkara, the ancient Memphis, which certainly date back for more than 5,000 years, show us the Egyptian type in its highest perfection, and with a more intellectual and I might say modern expression than is found 1,000 or 2,000 years later, when the type of the higher classes had evidently deteriorated somewhat from a slight infusion of African elements.

The same conclusion of the great distance at which any common point of divergence of the various races of mankind must be placed, is confirmed by a totally different line of inquiry, that into the origin of language.

Philologists have clearly proved that languages did not spring into existence ready made, like Minerva from the brain of Jupiter, but have followed the general law of Nature, and have had their periods of birth, growth, and evolution from simple into complex organism. Now there is a vast variety of languages, some say more than a thousand. A large proportion of these are, of course, only what may be called dialects of the same original language, as in the case of the whole Indo-European family, including Sanscrit, Zend, Greek, Latin, Teutonic, Celtic, and Slavonic, with all their offshoots and derived branches, as well as many others. These can be all traced back to the common root of the primitive language of an Aryan white race, who radiated by successive migrations from some region in the elevated plateaux of Central Asia. Any one who wants to be convinced of this has only to refer to Max Müller's works and trace the history of one verb, viz., that used to denote individual existence.

Asmi in Sanscrit has become *eimi* in Greek, *sum* in Latin (whence *sono, suis,* and all the modern derivatives of Latin races), and "am" in English; while the Latin *est*, the Greek *esti*, and the German *ist*, are clearly akin to the original *asti*. It may help in understanding how language has been formed if we point out that "I am" originally meant "I breathe," and "he is" is the more general and abstract form of "he stands."

But there are a number of languages between which no such relationship can be traced, which are constructed on radically different principles, and have no resemblance with one another in their roots, or primitive sounds used to express objects and simple ideas, except in the few cases where it can be traced to importation from abroad, or to imitation of naturally suggested sounds, such as those which have led so many nations to express the idea of "mother" by a sound resembling the bleating of a lamb. Obviously, similarity of sound in such words as are used for the ideas of father, mother, cow, crow, thunder, crack, splash, and so on, suggests no common origin, and as most, or at any rate a great many roots, were probably derived originally in this manner, though long since diverted to express other ideas by associations which it is impossible to trace, the wonder rather is that we should find so many languages with so few roots in common. The best authorities tell us that a list of fifty to one hundred languages could be made of which no one has been satisfactorily shown to be related to any other.

The main distinction between languages, however, is to be found in their inner mechanism, or grammar, rather than in the mere difference of root-sounds. The result of years of mechanical training in barbarous Latin and Greek grammars in our English public schools has been to leave the average Englishman completely ignorant of the real meaning of the word "grammar," and almost incapable of comprehending that it can mean anything else than a string of arbitrary rules to be learned by heart for the vexation of small boys.

And yet grammar is really most interesting, as showing the modes by which the dawning human intellect has proceeded, at remote periods and among different races, in working out the great problem of articulate speech, by which man rises into the higher regions of thought and is mainly distinguished from the brute creation. Consider first what the problem is, and then some of the principal modes which have been invented to solve it.

Suppose some primitive race to have accumulated a certain stock of root-words, or simple sounds to signify definite objects and simple ideas, they must soon find that these alone are not sufficient to convey briefly and clearly to other minds the ideas which they wish to express. For instance, suppose a tribe had got root words to express the ideas of "man," "bear," and "kill." What one of the tribe wants to convey from his own mind to that of his neighbour may be, "The man has killed the bear," or "The bear has killed the man," or "The" (or "A") man has killed a bear," or "bears," or "will" or "may have" killed, and so on through a vast number of variations on the original three-note theme. Up to a certain point, a man might succeed in making himself understood by using his three root-sounds in a certain order, aided by the pantomime of accent and gesture; and the Chinese, though one of the oldest civilised peoples of the world, have scarcely got beyond this stage. But the process would be difficult and uncertain, and at length it would occur to some genius that such modifications as those of definite and indefinite, past and present, singular and plural, etc., were of general application, not to the particular three or four roots which he wished to connect, but to all roots. The next step would be to invent a set of sounds which, attached in some way to the root-sounds, should convey to the hearer the sense in which it was intended that he should take them.

This is the fundamental idea of grammar, but it has been worked out by different races in the most different manner. The Chinese and other allied races in the South-east of Asia, such as the Burmese and Siamese, have solved it in the simplest manner. Their languages are what is called monosyllabic—that is, each word consists of a single syllable, and is a root expressing the fundamental idea, without distinction of noun from verb, active from passive, or other modifications. They have to trust, therefore, to express their meaning, mainly to syntax, or the order in which words succeed one another, which, up to a certain point, is the simplest method, and is largely adopted in modern English. Thus, "Man kill bear," "Bear kill man," convey the meaning just as clearly as the classical languages do by cases, when they distinguish whether the man is the killer or the killed by saying *homo* or *hominem*. But the monosyllabic system limits the nations who use it to an inconveniently small number of words, and fails in expressing their more complex relations, so that we find the same word in Chinese or Siamese often expressing the most different ideas, and the meaning can only be conveyed by supplementing the root-words and syntax by accent and other conventional signs which are akin to the primitive devices of gesture language. Thus, in Siamese, the syllable *ha*, according to the note in which it is intoned, may mean a pestilence, the number five, or the verb "to seek."

This very primitive and almost infantine form of language is confined to one family, that of the Chinese and Indo-Chinese, who, it may be observed, are by no means simple or primitive in other respects, but stand and have stood for centuries at a comparatively high level of civilisation. All other races, including the most savage, have adopted some form or other of grammar, *i.e.*, of modifying original root-sounds by additional generic sounds of definite determination; but the devices on which they have hit for this purpose are most various. Thus, the grammar of the Aryan family of languages has been formed by reasoning out such general categories of thought as articles, pronouns, and prepositions, coining sounds for them and prefixing these sounds to the root-sounds as separate determinating signs. More complex shades of meaning are conveyed principally by inflections, *i.e.*, by adding certain generic new sounds to the original root-word, and incorporating them with it so as to form modifications which are a sort of secondary words. Thus the ideas of present, past, and future love, loving, and being loved, lovely, and so on, are formed by transforming the root *amo* into such modifications as *amor, amavi, amabo, amans, amabilis*, etc. We can see this process in the course of formation in the change which converted the old English form "Cæsar his" into the modern genitive "Cæsar's."

Other families again obtain the same results by very different processes. The Semitic languages, for instance, including Hebrew, Arabic, Assyrian, and Phœnician, are what is called "triliteral," *i.e.*, they consist of roots mostly of three consonants, and express different shades of grammatical meaning by altering the internal vowels. Thus, from the root m-l-k are derived *melek*, a king; *malak*, he reigned, and so on.

The Turanian family, comprising Huns, Turks, Finns, Lapps, and other Mongolian races of Northern Asia, all speak agglutinative languages, *i.e.*, languages in which the root is put first and is followed by suffixes strung on to it, but not incorporated with it and remaining distinct. Thus in Turkish, the root *sev*, to love, is expanded into *sevishdirilmedeler*, meaning "incapable of being brought to love one another."

These are only given as specimens of some of the most marked of the vast varieties of language which have been examined and classified by philologists. They suggest a great many interesting reflections, but I confine myself to those which bear more immediately on the subject of man's origin and development. It is evident that they imply great antiquity for the existence, not of man only, but of separate races of men speaking separate languages.

Babylonian inscriptions, quite 4,000 years old, show that the characteristic features of the Aryan and Semitic languages were as clearly established then as they are now; and the hieroglyphics of Egyptian monuments, 1,000 years older, show the Coptic language essentially the same as modern Coptic, and although presenting some points of analogy with Semitic, too different to be classed with it. If these are descended from a common ancestor, clearly their origin must be extremely remote. And even with unlimited time it is difficult to conceive how such radical differences in the structure of languages could have arisen unless the different races had branched off before any clear form of articulate speech had become fixed. Could a race accustomed for generations to the free-flowing inflectional Aryan, have deserted it for the cramped forms of the Semitic, or *vice versâ*, could the Semite have adopted the modes of thought and expression of Sanscrit? And the same difficulty would apply in at least twenty or thirty cases of other families of language.

It must be recollected that language is not merely the conventional instrument of thought, but to a great extent its creator, and the mould in which it is cast. The mould may be broken, and races abandon old and adopt new languages by force of external circumstances, such as conquest or contact with and absorption by superior races, but there is no instance of its being so transformed from within as to pass into a totally different type. Nor can we very well see how rootwords once attached to fundamental ideas, such for instance as the simpler numerals, should come to be forgotten and new and totally different words invented.

Of course, the explanation was easy in the olden days, when everything was referred to miracle. Languages were different because God had made them so, to baffle the attempt of united mankind to build a tower high enough to reach to heaven. But the theory of special miraculous creation for each language cannot stand a moment's investigation.

As in the case of the animal world, special creations, if admitted at all, must be multiplied to an extent which becomes absurd. Is every petty tribe of savages who speak a language unintelligible to others to be supposed to have had it conferred upon it as a miraculous gift? Was the language of the extinct Brazilian tribe, of which Humboldt tells us that a very old parrot spoke the last surviving words, one of the languages used to scatter the builders of the Tower of Babel? Or, still more conclusively, where we know and can prove that one part of a language is the product of natural laws, can we assume that another part of the same language is the result of miracle? Did it require Divine inspiration to make the old Egyptians call a cat *miaou*, or to teach so many nations to express the idea of mother by imitating the bleating of a lamb? If not, why should half the words in a dictionary be miraculous and half natural?

And if Cæsar is correctly reported to have been more proud of discovering a new case than of conquering Gaul, ought we not to "render unto Cæsar the things that are Cæsar's," and assign grammar as well as words to human invention? In short, no reasonable man who studies the subject can doubt that language is just as much a machine of human invention for communicating thought, as the spinning jenny is for spinning cotton.

The general conclusion, then, to be drawn from the study of language points in the same direction as that of all other branches of science, viz., that their true history is that of evolution from simple origins by the operation of natural laws over long periods of time into forms of greater complexity and higher development. What language really does for us is to take up the thread where the oldest history fails us, and show that even at this date it is impossible to doubt that the human race must have been already in existence for a very long period, and in existence as at the present day in several sharply distinguished varieties, so that the common origin, if there be one, must be placed still further back. As history verified by the Egyptian monuments extends over a period of nearly 7,000 years, this is equivalent to saying that such a period can only be a very small part of the total time which has elapsed since man became an inhabitant of the earth.

The origin and development of religions have been much discussed, but too often with a desire to make theories square with wishes. The subject also does not admit of such precise determination as in treating of arts and languages, which have left traces of themselves in the form of primitive implements and primitive roots.

The history of religions really begins with written records, or at the earliest with the older myths which are embodied in these records. But these are all comparatively modern, and imply a considerable progress in civilisation before they could have existed. If we wish to form some idea of what may have been the primitive elements from which religion was evolved, during the long Neolithic and still longer Palæolithic periods which preceded history, we must look at what are actually the religious ideas of contemporary savage and semi-barbarous races.

At the very lowest stage of savagery we find races like the Australians, the Bushmen, the Mincopies, and the Fuegians, who cannot be said to have any religion at all, or at the most some vague ideas of ghosts and spirits. The Mincopies of the Andaman Islands, who are considered by Professor Owen as "perhaps the most primitive, or lowest in the scale of civilisation, of the human race," are reported by Dr. Mowatt to have "no idea of a Supreme Being, no religion, nor any belief in a future state of existence." Sir J. Lubbock says of the Australians that "they have no religion, nor any idea of prayer; but most of them believe in evil spirits, and all have great dread of witchcraft."

As we rise above this level of the lowest savagery we find ideas of religion beginning to grow from two main tap-roots. The first is the idea of ghosts or spirits, which arises naturally from dreams and visions and develops itself into ancestor and hero-worship, and belief in a world of spirits, good and evil, influencing men's lives and fortunes, and in many forms of sickness taking possession of their bodies. This spirit-worship also necessarily leads to some dim perception of a future life.

The other tap-root is the inevitable disposition to account for the phenomena of nature, when men first began to reflect on them, by the agency of invisible beings like themselves; in other words, of anthropomorphic gods. This is a higher and later stage of religious belief than the former, for it implies a certain disposition to inquire into the causes of things and a certain amount of reasoning power to infer like causes from like results.

But the two often blend together, as in the religions of the Aryan race, in which we see deified heroes and ancestors crowding the courts of Olympus, with a multitude of anthropomorphic gods, who are often merely obvious personifications of natural phenomena or astronomical myths. Thus Varuna, Ouranos, or Uranus, are personifications of the vault of heaven; Phœbus, the shining one, of the sun; Aurora, of the dawn; while Hercules is half deified hero and half solar myth. Sometimes, however, of the two stems of religion one only has flourished, and the other has either never existed, or been overshadowed by the first and relegated to a lower sphere. Thus the great Chinese civilisation, comprising such a large portion of the human race, has apparently developed its religion entirely from the idea of spirits and spirit-worship. The worship of ancestors is its main feature, and its sacred books are, in effect, treatises on ethics and political economy, with rules for rites and ceremonies to enforce decent and decorous behaviour, rather than what we should call works of religion. There is no trace of a conception of anthropomorphic gods in the genuine national Chinese religion from Confucius downwards; and even the introduction of Buddhism has done little but add the deified hero, Buddha, to the list of divine ancestors and give more definite shape to various vague superstitions. In like manner the whole Buddhist world can hardly be said to recognise anything beyond their incarnate hero, except a Nirvana or metaphysical abstraction, rather than a personal deity.

With other races again, and specially the Hebrew, the idea of a tribal anthropomorphic God has gradually swallowed up that of other gods, developed into that of one Almighty Being, and dwarfed that of ghosts and spirits. The primitive Hebrews, indeed, carried this so far as to exclude all ideas of a future life from their religious system. Their primitive God, however, was strictly anthropomorphic, and modelled on the idea of an Oriental sultan—sometimes good and beneficent, but sometimes cruel and capricious, and above all jealous of any disrespect and enraged by any disobedience. Morality seems at first to have had little or nothing to do with these conceptions, and there is not the remotest trace in the early history of any religion, of its having been born ready-made from the necessary intuition of one Almighty God of love, mercy, and justice, which is so confidently assumed by many metaphysicians and theologians. On the contrary, conscience had to be first evolved, and the process may be followed step by step by which, as manners became milder and ideas purer, the grosser attributes of Deity were gradually purged off, and the idea of a just and merciful God was evolved from barbaric elements.

These considerations, however, lead us far from the question of the first dawn of religion among primitive man. Judging from the earliest facts of history, and the analogy of modern savage races, where we might look for the first traces of religious ideas would be from the contents of tombs and from idols. When a tribe had attained to some definite idea of a future life it would almost certainly bury weapons and implements with its dead, as is the case with modern savages. When it had reached the stage of worshipping anthropomorphic deities, it would probably frame images of them, some of which would be found in their tombs and dwellings.

The latter test soon fails us. In the early Egyptian tombs, and in the remains of the prehistoric cities excavated by Dr. Schliemann, images of owl and ox-headed goddesses, and other symbolical figures or idols, are found in abundance. But when we ascend into Neolithic times, such idols are no longer found, or, if found, it is so rarely that archæologists still dispute as to their existence. Certain crescents found in the Swiss lake-dwellings were at one time thought to indicate a worship of the moon, but the better opinion seems to be that they were used as rests for the head during sleep, as we find similar objects now used in many parts of the world. Among the many thousand objects recovered from these Swiss lake-dwellings and other Neolithic abodes, there are only a very few which may possibly have been rude idols or amulets, and the only ones which may be said with some certainty to have been idols, are one or two discovered by Mons. de Braye in some artificial caves of the Neolithic period, excavated in the chalk of Champagne, which appear to be intended for female figures of life size with heads somewhat resembling that of the owl-headed Minerva.

When we pass to Palæolithic times the evidence of idols becomes more faint, and rests solely on the conjecture that some of the figures carved by the Reindeer-men of La Madeleine and other caves, may probably have been intended for amulets. As they were such skilful carvers, and so fond of drawing whatever impressed itself on their imagination, the presumption is strong that they had not advanced to the stage when the worship of gods symbolised by idols had come into existence, as otherwise more undoubted idols must have been found in the caves which were so long their habitations, and which have yielded such a number of remains of works of art.

The evidence for a belief in a future existence and in spirits is more conclusive. Throughout the whole Neolithic period we find objects buried with the dead which were evidently intended for use in a future life. We find also in many Neolithic tombs a singular fact which points to the existence of a very long belief in evil spirits. Many of the skulls, especially of young people, have been trepanned, that is, a piece of the skull has been cut out, making a hole, apparently to let out the evil spirit which was supposed to be causing epilepsy or convulsions; and where the patient had recovered and the wound healed, when he died long afterwards, a piece of the skull, including this trepanned portion, was sometimes cut out and used apparently as an amulet. The objects deposited in graves show that the idea of a future life was, as with most savages of the present day, that of a continuation of the same life as he had led here, though perhaps in happier hunting-grounds. In some cases a great chief seems to have had wives and slaves slaughtered and buried with him, though the proofs of this are more clear and abundant in later prehistoric times than during the Neolithic period. Cannibalism, however, seems to have occasionally prevailed both in Palæolithic, Neolithic, and prehistoric times, as it did so extensively among modern savage races before they came under civilising influences. This is clearly proved by the number of human bones, chiefly of women and young persons, which have been found charred by fire and split open for extraction of the marrow.

The evidence of belief in a future life becomes more rare and uncertain in Palæolithic times. Perhaps it may be because we have so few authentic discoveries of Palæolithic burying-places, and so many instances of caves, once inhabited by Palæolithic races, being used long afterwards as Neolithic sepulchres. After the famous cave of Aurignac it is difficult to trust any evidence of the discovery of a real Palæolithic sepulchre which has not been subsequently disturbed.

In the few cases also where Palæolithic skeletons have been found, as in that of the men of Neanderthal and Mentone, they have often been those of single individuals, and it may be doubted whether they were buried there, or merely died in the caves in which they lived, in which case any implements found with them do not necessarily imply that they were placed there for use in a future life. On the whole it seems doubtful whether any certain proofs of burials denoting knowledge of a future life can be found in Palæolithic times, and if there are, they are certainly few and far between, and confined to the later stages of that period.

All we can say is, that religion certainly did not descend ready-made among these aboriginal savages, but that, like language, it was slowly developed from beginnings as rude as those we now find among the lowest races of savages.

It may be well, however, to say here, once for all, what is applicable to many other passages in this book, that the question of the origin of any religion is entirely different from that of its truth or falsehood. To explain a thing is not to disprove it; on the contrary, a thing only really becomes true to us when we understand it. A stately oak, with wide-spreading branches, that give shade and shelter to the cattle of the fields, is not the less a fact because we know that it did not drop ready-made from heaven, but grew from an acorn. The intrinsic truth of a religion must be tested by the conformity which, in a given stage of its evolution, it bears to the facts of the universe as disclosed by science, and to the feelings and moral perceptions which have been equally developed by evolution in the contemporary world.

All I contend for is, that all religions have grown and been developed from humble origins, and that their history, impartially considered, does not contradict, but on the contrary greatly confirms the law of natural evolution.

Of the two faculties by which man is commonly distinguished from the brute creation, viz., that of being the speaking and the tool-making animal, the former attribute has been shown to be the product of evolution from origins long since lost in the far-off distance of remote ages.

The same remark is even more certainly true as regards the other attribute of tool-making, or, in its widest sense, adapting natural laws and natural objects to the arts of life by intelligent application. The primitive roots, so to speak, of this industrial language, which in the case of spoken language for the most part elude our search, are here furnished by the Palæolithic remains found so abundantly in river drifts and caves. There can be no doubt whatever that the modern woodcutter's axe and carpenter's adze are the lineal descendants of the rudely-chipped *hâches*, or celts, which are dug out of the gravels of St. Acheul, or from below the stalagmite of Kent's Cavern. The regular progression can be traced from the mass of flint rudely chipped to a point, with a butt-end left rough to grasp in the hand, up to more symmetrical and carefully-chipped forms;

to implements intended to be hafted or fastened to a handle; to implements ground and polished to a sharp edge and pierced for the handle; and finally to the finished specimens of the later Neolithic period, which exactly represent the adze and battle-axe, and are almost identical with those used quite recently by the Polynesians and other semi-civilised races who had no access to metals. From these the transition to metals is easily traced, the first bronze implements and weapons being facsimiles of those of polished stone which they superseded, and the gradual development of bronze, and from bronze to the cheaper and more generally useful metal, iron, being a matter of quite modern history.

In like manner, the development of the knife, sword, and all cutting instruments, from the primitive flint-flake, can be traced step by step, and is beyond doubt; and equally so the development of all missiles, from the primitive chipped flint, used as a javelin or arrow-head, up to the modern rifle. When we catch the first glimpse of the beginnings of human art or industry, the furniture or stock-in-trade of Palæolithic man appears to have been as follows :

He was acquainted with fire. This seems to be clearly established by the charred bones, charcoal, and other traces of fire which are found in the oldest Palæolithic caves, and even in the far distant Miocene period, if we can believe in the flints discovered by the Abbé Bourgeois in the strata of Thenay, some of which appear to have been split by the action of fire. This is a remarkable fact, for a knowledge of the means of kindling fire is by no means a very simple or obvious attainment. Apes and monkeys will sit before a fire and enjoy its warmth, but no monkey has yet developed intelligence enough even to put fresh sticks on to keep

up the fire, much less to rekindle it when extinct.
Primeval man must often have had experience of
fire from natural causes, as from forests and prairies
scorched by a tropical sun being set on fire by lightning, or from volcanic eruptions ; but how he learned
from these to kindle fire for himself is not so obvious. Savage races, as a rule, do so by converting
mechanical energy into heat, by the friction of a
stick twirled round in a hole, or rubbed backwards
and forwards in a groove in another piece of wood ;
and there are old observances among civilised nations
which show that this was the mode practised by
their ancestors, as when the sacred fire in the Temple
of Vesta was relighted in this manner by the old
Romans if it had chanced to be extinguished. It is
probable, therefore, that this was the original mode of
obtaining fire, but if so, it must have required a good
deal of intelligence and observation, for the discovery
is by no means an obvious one, nor is it easy to see any
natural process that might suggest it.

Neither ancient history nor the accounts of existing
savage races throw much light on the question. The
narratives of the discovery of fire contained in the
oldest records are obviously mythical, like the fable of
Prometheus, which is itself a version of the older Vedic
myth of the god Agni (whence the Latin *ignis* or fire)
having been taken from a casket and given to the first
man, Manou, by Pramantha, which in the old Vedic
language means taking forcibly by means of friction. Of
the same character are the mythical legends of savage
races of fire having been first brought by some wonderful bird or animal ; and there is nowhere anything like
an authentic tradition of the fact of its first introduction There have been reports of savages who were

unacquainted with fire, but they have never been well
authenticated, and the nearest approach to such a state
of things was probably furnished by the aborigines of
Van Diemen's Land, of whom it is said that in all their
wanderings they were particularly careful to bear in
their hands the materials for kindling a fire, in the
shape of a firebrand, which it was the duty of the
women to carry, and to keep carefully refreshed from
time to time as it became dull.

On the whole, traditions all point to fire having
been first obtained from friction, and it is possible that
the first idea may have been derived from the boughs of
trees, or silicious stalks of bamboos, having been set on
fire when rubbed together by the action of the wind.

It is easier to see the origin of the remaining equipment of primitive man, viz., chipped stones, for flints
splintered by frost or fire often take naturally the
forms of sharp-edged flakes and rude hatchets or
hammers, and very little invention was required to
improve these specimens, or endeavour to imitate them
by artificial chippings. It is rather surprising that this
art did not improve more rapidly, for it is evident that
the old Palæolithic period must have lasted a long time
before any decided progress began to show itself. And
during this long period a singular uniformity appears to
have prevailed throughout the Palæolithic world. The
rude form of the celt or *hâche*, with a blunt butt
and chipped roughly to a point, is found in the oldest
river gravels and caves wherever they have been in-
vestigated, and the forms of the Somme and the Thames
are repeated in the quartzite implements of the Madras
laterite.

In the very oldest caves and river deposits the tool-
equipment of man seems to have been very much limited

to these rude celts, used probably for smashing skulls in war and the chase, and splitting bones to get at the marrow; sharp-edged flakes for cutting; rude javelin-heads; and stones chipped to a rounded edge, very like those used by the Esquimaux for scraping bones and skins. As we ascend in time we find arrow-heads of stone and bone, at first unbarbed and gradually becoming barbed, showing that the bow had been discovered; harpoons of bone and fish-hooks; bone pins and needles; and a much greater variety and more carefully-chipped forms of flint tools and weapons; until we finally reach the upper reindeer stage of caves like that of La Madeleine, where artistic drawings and carvings are found, and the equipment generally is superior to that of many existing savage tribes, and not much inferior to that of the Esquimaux and other Arctic races.

We then pass into Neolithic times, when many of the chief elements of civilisation are already in full force. Man has emerged in many localities from the hunter into the pastoral stage, the principal domestic animals are known, and in some of the later lake-dwellings he has advanced a stage further, and has become an agriculturist living in villages. From this to the Bronze and early historical periods, there is no great break, and the ruder tribes of barbarians described by Cæsar and Tacitus may well have been the lineal descendants of the Neolithic men whose polished axes and finely-shaped arrow-heads lie scattered over the surface of Europe and are found in innumerable burial-mounds and dolmens.

But in Palæolithic times, though we can see constant progress, mankind is still in a state of unmitigated barbarism. Agriculture was clearly unknown, for the hand-mills, pestles, and mortars, which are

DEVELOPMENT OF THE ARROW.

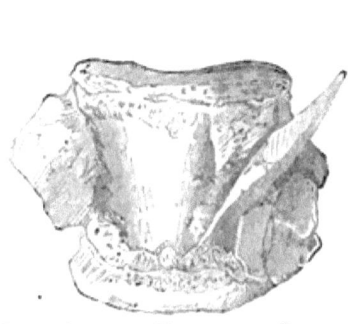

FLINT ARROW IN VERTEBRA OF REINDEER.
Palæolithic. La Madeleine.

PALÆOLITHIC.
Mammoth Period. Le Moustier

PALÆOLITHIC.
Reindeer Period.
First vestige of barb.

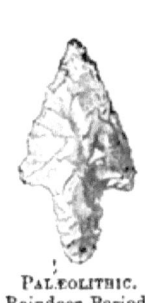

PALÆOLITHIC.
Reindeer Period.

PALÆOLITHIC.
Reindeer Period.

NEOLITHIC.
Denmark.

NEOLITHIC.
Ireland.

NEOLITHIC.
Denmark.

RECENT.
Esquimaux.

(From Lubbock's "Prehistoric Times.")

among the most enduring and abundant relics where grain was used for food, are never met with. Pottery was unknown in all the earlier periods, and it is questionable whether even the rudest forms of baked clay, moulded by hand, are found where there is no intermixture of a subsequent Neolithic habitation. The dog was clearly not a companion of man prior to the era of the Danish kitchen-middens, for the spongy parts of bones which are always gnawed by dogs when dogs are present, are invariably preserved in the *débris* of Palæolithic caves, and the few bones of dogs, wolves, and foxes found with human remains in these caves almost always show that the animals had formed part of the food of the inhabitants.

Other domestic animals were, in all probability, equally unknown, although it has been thought possible that some of the tribes of the reindeer period may have had herds of the half-tame deer, like the modern Laplanders. This conjecture, however, appears to rest solely on the large number of bones and horns found at certain stations, which may have arisen from their having been occupied for a very long period, and as the dog was unknown, it seems probable that no other animals had been domesticated.

As regards clothing, the first certain proofs of its use are afforded by the bone pins and needles, which were evidently employed for fastening the skins of animals together, and the scrapers were probably used for scraping these skins and fashioning the bone implements. It is probable, therefore, that the use of skins as a protection against the cold of the Glacial period, was known at a very early period.

Ornaments, also, are of very early date, as pierced shells, sometimes fossil, and pierced teeth of the bear and other animals are frequently found under circumstances which show that they must have been strung together as necklaces. The skeleton found in a cave at Mentone had a number of perforated shells of Nassa, and a few stag's teeth also perforated, dispersed about the skull, so as to show that they had formed some sort of head ornament. Lumps of red hematite, also, probably used for paint, have been found in some of the caves of the reindeer period.

Captain Cook's description of the savages of Tierra del Fuego would have applied to them, that, "although content to be naked, they were very ambitious to be fine;" and probably like these poor Fuegians, they adorned themselves with streaks of red, black, and white, and wore bracelets and anklets of shell and bone.

If we wish to form some idea of the manners and customs of our Palæolithic ancestors, we must look for them among the existing savage races, whose mode of life, and equipment of tools and weapons, most nearly resemble those of the earliest cave-dwellers. The Australians, the Bushmen of South Africa, the Mincopies of the Andaman Islands, and the Fuegians are probably the lowest specimens of the human race known in modern times; but even these are in some respects further advanced in the arts than the first Palæolithic man. The Bushmen are skilled in the use of the bow, and have discovered the art of poisoning their arrows. The Australians, Mincopies, and Fuegians have canoes, harpoons, and fish-hooks. The latter approach more nearly to the conditions of life of the savages who accumulated the kitchen-middens on the coasts of Denmark at a much later period, and the Bushmen probably represent better those of the cave-men who lived principally on the produce of the chase of large animals, such as the mammoth, rhinoceros, cave bear, horse, and deer. The pigmy Bushman will attack the elephant, the rhinoceros, and even the lion, and often succeed in killing them by pitfalls or poisoned arrows.

The inferences, therefore, to be drawn, alike from the physical development of the individual man, and from the origin and growth of all the faculties which specially distinguish him from the brute creation— language, religion, arts, and science—all point to the conclusion that he is a product of laws of evolution, and not of special or miraculous creation.

Still, admitting this, we must admit on the other hand, that until more of the " missing links " are discovered, and the origin of man is placed on a basis of scientific certainty, there is an opening left for the belief that here, if nowhere else, there was some supernatural interference with the laws of Nature, and that the finger of the clock-maker did here alter the hands of the clock from the position which they would have occupied under the original law of its construction. But if this were so, it must equally in candour be admitted that the miracle did not consist in placing man and woman upon earth, at any recent period, or with faculties in any way developed, but could only have consisted in causing a germ or germs to come into existence, different from any that could have been formed by natural evolution, and containing within them the possibilities of conscious and civilised man, to be developed from the rudest origins by slow and painful progress over countless ages.

Part II.

MODERN THOUGHT.

CHAPTER VII.

MODERN THOUGHT.

Lines from Tennyson—The Gospel of Modern Thought—Change exemplified by Carlyle, Renan, and George Eliot—Science becoming Universal—Attitude of Orthodox Writers—Origin of Evil—First Cause unknowable—New Philosophies and Religions—Herbert Spencer and Agnosticism—Comte and Positivism—Pessimism—Mormonism—Spiritualism—Dreams and Visions—Somnambulism—Mesmerism—Great Modern Thinkers—Carlyle—Hero-worship.

LIV.

Oh yet we trust that somehow good
   Will be the final goal of ill,
   To pangs of nature, sins of will,
Defects of doubt, and taints of blood;

That nothing walks with aimless feet;
   That not one life shall be destroy'd,
   Or cast as rubbish to the void,
When God hath made the pile complete;

That not a worm is cloven in vain;
   That not a moth with vain desire
   Is shrivel'd in a fruitless fire,
Or but subserves another's gain.

Behold, we know not anything.
   I can but trust that good shall fall
   At last—far off—at last, to all,
And every winter change to spring.

So runs my dream: but what am I?
  An infant crying in the night:
  An infant crying for the light:
And with no language but a cry.

LV.

The wish, that of the living whole
  No life may fail beyond the grave,
  Derives it not from what we have
The likest God within the soul?

Are God and Nature then at strife,
  That Nature lends such evil dreams
  So careful of the type she seems,
So careless of the single life;

That I, considering everywhere
  Her secret meaning in her deeds,
  And finding that of fifty seeds
She often brings but one to bear,

I falter where I firmly trod,
  And falling with my weight of cares
  Upon the great world's altar-stairs
That slope thro' darkness up to God,

I stretch lame hands of faith, and grope,
  And gather dust and chaff, and call
  To what I feel is Lord of all,
And faintly trust the larger hope.

LVI.

"So careful of the type?" but no.
  From scarped cliff and quarried stone
  She cries, "A thousand types are gone:
I care for nothing, all shall go.

"Thou makest thine appeal to me:
  I bring to life, I bring to death:
  The spirit does but mean the breath:
I know no more." And he, shall he,

> Man, her last work, who seem'd so fair,
>   Such splendid purpose in his eyes,
>   Who roll'd the psalm to wintry skies,
> Who built him fanes of fruitless prayer,
>
> Who trusted God was love indeed,
>   And love Creation's final law—
>   Tho' Nature, red in tooth and claw
> With ravine, shriek'd against his creed—
>
> Who loved, who suffer'd countless ills,
>   Who battled for the True, the Just,
>   Be blown about the desert dust,
> Or seal'd within the iron hills?
>
> No more? A monster then, a dream,
>   A discord. Dragons of the prime,
>   That tare each other in their slime,
> Were mellow music match'd with him.
>
> O life as futile, then, as frail!
>   O for thy voice to soothe and bless!
>   What hope of answer, or redress?
> Behind the veil, behind the veil.
>
>   TENNYSON, *In Memoriam*.
>   (*By kind permission of* LORD TENNYSON.)

THESE noble and solemn lines of a great poet sum up in a few words what may be called "the Gospel of Modern Thought." They describe what is the real attitude of most of the thinking and earnest minds of the present generation. On the one hand, the discoveries of science have so far established the universality of law, as to make it impossible for sincere men to retain the faith of their ancestors in dogmas and miracles. On the other, larger views of man and of history have shown that religious sentiment is an essential element of human nature, and that many of our best feelings, such as love, hope, conscience, and reverence, will always seek to find reflections of themselves in the unseen world. Hence faith has diminished and charity increased. Fewer believe old creeds, and those who do, believe more faintly; while fewer denounce them, and are insensible to the good they have done in the past and the truth and beauty of the essential ideas that underlie them.

On the Continent, and especially in Catholic countries, where religion interferes more with politics and social life, there is still a large amount of active hostility to it, as shown by the massacre of priests by the French Communists; but, in this country, the old Voltairean infidelity has died out, and no one of ordinary culture thinks of denouncing Christianity as an invention of priestcraft. On the contrary, many of our leading minds are at the same time sceptical and religious, and exemplify the truth of another profound saying of Tennyson:

> There is more faith in honest doubt,
> Believe me, than in half the creeds.

The change which has come over modern thought cannot be better exemplified than by taking the instance of three great writers whose works have produced a powerful influence—Carlyle, Renan, and George Eliot. They were all three born and brought up in the very heart of different phases of the old beliefs—Carlyle, in a family which might be taken as a type of the best qualities of Scottish Presbyterianism, bred in a West country farmhouse, under the eye of a father and mother whom he loved and revered, who might have been the originals of Burns' "Cotter's Saturday Night," or the descendants of the martyrs of Claverhouse. His own temperament strongly inclined to a stern Puritanical piety; his favourite heroes were Cromwell and John Knox; his whole nature was antipathetic to science. As his biographer, Froude, reports of him, "He liked ill men like Humboldt, Laplace, and the author of the 'Vestiges.' He refused Darwin's transmutation of species as unproved; he fought against it, though I could see he dreaded that it might turn out true." And yet the deliberate conclusion at which he arrived was that "He did not think it possible that educated honest men could even profess much longer to believe in historical Christianity."

The case of Renan was equally remarkable. He was born in the cottage of Breton peasants of the purest type of simple, pious, Catholic faith. Their one idea of rising above the life of a peasant was to become a priest, and their great ambition for their boy was that he might be so far honoured as one day to become a country curé. Young Renan, accordingly, from the first day he showed cleverness, and got to the top of his class in the village school, was destined for the priesthood. He was taken in hand by priests, and found in them his kindest friends; they sent him to college, and in due time to the Central Seminary where young men were trained for orders. All his traditions, all his affections, all his interests, led in that direction, and yet he gave up everything rather than subscribe to what he no longer believed to be true. His conversion was brought about in this way. Having been appointed assistant to a professor of Hebrew he became a profound scholar in Oriental languages; this led to his studying the Scriptures carefully in the original, and the conclusion forced itself upon him that the miraculous part of the narrative had no historical foundation. Like Carlyle, the turn of his mind was not scientific, and while denying miracles he remained keenly appreciative of all that was beautiful and poetical in the life and teaching of Jesus, which he has brought more vividly before the world in his writings than had ever been done by orthodox commentators.

George Eliot, again, was brought up in yet another phase of orthodox Christianity—that of middle-class nonconformist Evangelicalism. She embraced this creed fervently, and, as we see in her "Dinah," retained a keen appreciation of all its best elements. But as her intellect expanded and her knowledge widened, she too found it impossible to rest in the old belief, and, with a painful wrench from a revered father and loving friends, she also passed over from the ranks of orthodoxy. She also, after a life of profound and earnest thought, came to the conclusion recorded of her by an intimate friend and admirer, Mr. Myers:

"I remember how at Cambridge, I walked with her once in the Fellows' Garden of Trinity, on an evening of rainy May; and she, stirred somewhat beyond her wont, and taking as her text the three words which have been used so often as the inspiring trumpet-calls of men—the words *God*, *Immortality*, *Duty*—pronounced, with terrible earnestness, how inconceivable was the *first*, how unbelievable the *second*, and yet how peremptory and absolute the *third*. Never, perhaps, had sterner accents affirmed the sovereignty of impersonal and unrecompensing law. I listened, and night fell; her grave, majestic countenance turned toward me like a Sibyl's in the gloom; it was as though she withdrew from my grasp, one by one, the two scrolls of promise, and left me the third scroll only, awful with inevitable fates."

Such instances as these cannot be the result of mere accident. As long as scepticism was confined to a limited number of scientific men it might be possible to think that it was merely the exaggeration of a particular train of thought pursued too exclusively. But when science has become the prevailing mode of thought, and has been brought home to the minds of all educated persons, it is no longer possible to represent it as an exceptional aberration. And where the bell-wethers of thought lead the way the flock will follow. What the greatest thinkers think to-day, the mass of thinkers will think to-morrow, and the great army of non-thinkers will assume to be self-evident the day after. This is very nearly the case at the present day; the great thinkers have gone before, the mass of thinkers have followed, and the still greater mass of non-thinkers are wavering and about to follow. It is no longer, with those who think at all, a question of absolute faith against absolute disbelief, but of the more or less shade of "faintness" with which they cling to the "larger hope."

This is nowhere more apparent than in the writings of those who attempt to stem the tide which sets so strongly against orthodoxy. They resolve themselves mainly into one long wail of "oh the pity of it, the pity of it!" if the simple faith of olden times should disappear from the world. They show eloquently and conclusively that science and philosophy cannot satisfy the aspirations or afford the consolations of religion. They expose the hollowness of the substitutes which have been proposed, such as the worship of the unknowable, or the cult of humanity. They win an easy triumph over the exaggerations of those who resolve all the historical records of Christianity into myths or fabulous fulfilment of prophecies, and they wage fierce battles over minor points, as whether the first quotations from the Gospels are met with in the first or second half of the second century. But they nowhere attempt to grapple with the real difficulties, and show that the facts and arguments which converted men like Carlyle and Renan are mistaken facts and unsound arguments. Attempts to harmonise the Gospels and to prove the inspiration of writings which contain manifest errors and contradictions, have gone the way of Buckland's proof of a universal deluge, and of Hugh Miller's attempt to reconcile Noah's ark and the Genesis account of creation with the facts of geology and astronomy. Not an inch of ground that has been conquered by science has ever been reconquered in fair fight by theology.

This great scientific movement is of comparatively recent date. Darwin's "Origin of Species" was only published in 1859, and his views as to evolution, development, natural selection, and the prevalence of universal law, have already annexed nearly the whole world of modern thought and become the foundation of all philosophical speculation and scientific inquiry.

Not only has faith been shaken in the supernatural as a direct and immediate agent in the phenomena of the worlds of matter and of life, but the demonstration of the "struggle for life" and "survival of the fittest" has raised anew, and with vastly augmented force, those questions as to the moral constitution of the universe and the origin of evil, which have so long exercised the highest minds. Is it true that "love" is "Creation's final law," when we find this enormous and apparently prodigal waste of life going on; these cruel internecine battles between individuals and species in the struggle for existence; this cynical indifference of Nature to suffering? There are, approximately, 3,600 millions of deaths of human beings in every century, of whom at least 20 per cent., or 720 millions, die before they have attained to clear self-consciousness and conscience. What becomes of them? Why were they born? Are they Nature's failures, and "cast as rubbish to the void?"

To such questions there is no answer. We are obliged to admit that as the material universe is not, as we once fancied, measured by our standards and regulated at every turn by an intelligence resembling ours; so neither is the moral universe to be explained by simply magnifying our own moral ideas, and explaining everything by the action of a Being who does what we should have done in his place. If we insist on this anthropomorphic conception we are driven to this dilemma. Carlyle bases his belief in a God, "the infinite Good One," on this argument: "All that is good, generous, wise, right—whatever I deliberately and for ever love in others and myself, who or what could by any possibility have given it to me but One who first had it to give? This is not logic; this is axiom."

But how of the evil? No sincere man looking into the depths of his own soul, or at the facts of the world around, can doubt that along with much that is good, generous, wise, and right, there is much that is bad, base, foolish, and wrong. If logic compels us to receive as an axiom a good author for the former, does not the same logic equally compel us to accept the axiom that the author of the latter must have been one who "first had it in himself to give"? That is, we must accept the theory of a God who is half good, half evil; or adopt the Zoroastrian conception of a universe contested by an Ormuzd and Ahriman, a good and evil principle, whose power is, for the present at any rate, equally balanced.

From this dilemma there is no escape, unless we give up altogether the idea of an anthropomorphic deity, and adopt frankly the scientific idea of a First Cause, inscrutable and past finding out; and of a universe whose laws we can trace, but of whose real essence we know nothing, and can only suspect or faintly discern a fundamental law which may make the polarity of good and evil a necessary condition of existence. This is a more sublime as well as more rational belief than the old orthodox conception; but there is no doubt that it requires more strength of mind to embrace it, and that it appears cold and cheerless to those who have been accustomed to see special providences in every ordinary occurrence, and to fancy themselves the special objects of supernatural supervision in all the details of daily life. Hopes and fancies, however, are powerless against facts; and the world is as surely passing from the phase of orthodox into that of scientific belief as youth is passing into manhood, and the planet which we inhabit from the fluid and fiery state into that of temperate heat, progressive cooling, and final extinction as the abode of life. In the meantime, what can we do but possess our souls in patience, follow truth wherever it leads us, and trust, as Tennyson advises, that in the long run everything will be for the best, and "every winter turn to spring"?

The decay of old religious beliefs, and the introduction of new conceptions based on scientific discovery, have given rise to many attempts to found new philosophies, and in some cases new sects and religions, of some of the principal of which a short account may be given.

One of the greatest thinkers of modern times, Herbert Spencer, has expanded the theories of modern science, specially those of the conservation of energy and of Darwinian evolution, into a generalised philosophy, embracing not only the phenomena of the material and living universe, but also history, religion, politics, and all the complex relations of social life. He starts from the principle that throughout the universe, in general and in detail, there is an unceasing redistribution of matter and motion. This shows itself as evolution where there is a predominant aggregation of matter and diminution of motion, and as dissolution where matter is disintegrated and motion increased. Thus, in the formation of coal, the motion of the sun's rays is fixed in the condensed matter of the chemical products of vegetation, and is dissipated when, after countless ages, the coal is burned and its substance dissolved into its elements. These changes constitute a transformation of the uniform or homogeneous into the differentiated or heterogeneous, as seen in the condensation of nebulous or cosmic matter into suns and planets; in the varied elements of the inorganic world; "in each organism, vegetable or animal; in the aggregate of organisms, thought and geologic time; in the mind; in society; in all products of social activity." These changes are all in the direction of passing from an indefinite whole to definite parts, and they are inevitable, unless the original substance were so absolutely uniform as to be absolutely stable.

Once started, this process of differentiation tends necessarily to go on, the surrounding conditions being ever at work, whether by aggregation or dissolution, by joining like to like, or separating unlike from unlike, to sharpen and make more definite existing differences.

This is in effect a generalised conception of Darwin's laws of the "struggle for life" and "survival of the fittest." Finally, however, the result of all these changes is that an ultimate equilibrium is reached, which is rest in the inorganic and death in the organic world; as when the sun with all its planets shall have parted with all its heat, and all its energy shall have run down to one uniform level. From this state it can only be roused by some fresh shock from without, dissipating it again into a mass of diffused matter and unbalanced motions.

Hence we come to the final statements of the Spencerian philosophy, as given in the words of its author:

"This rhythm of evolution and dissolution, completing itself during short periods in small aggregates, and in the vast aggregates distributed through space completing itself in periods which are immeasurable by human thought, is, so far as we can see, universal and eternal, each alternating phase of the process predominating, now in this region of space and now in that, as local conditions determine. All these phenomena, from their great features even to their minutest details, are necessary results of the persistence of force under its forms of matter and motion. Given these as distributed through space, and their quantities being unchangeable either by increase or decrease, there inevitably result the continuous redistributions distinguishable as evolution and dissolution, as well as those special traits above enumerated. That which persists, unchanging in quantity, but ever changing in form, under these sensible appearances which the universe presents to us, transcends human knowledge and conception, is an unknown and unknowable power, which we are obliged to recognise as without limit in space and without beginning or end in time."

This is, in its highest form, the philosophy of Agnosticism. A very different thing, be it observed, from Atheism, for it distinctly recognises an underlying power which, although "unknown and unknowable," may be anything harmonising with the feelings and aspirations in which all religious sentiment has its origin, so long as it fulfils the condition of not, by too precise definition, coming into collision with something which is not "unknown" but "known" and irreconcilable with it.

For instance, there is nothing in Agnosticism to negative the possibility of a future state of existence. Behind the veil there may be anything, and no one can say that individual consciousness may not remain or be restored after death, and that our condition may not be in some way better or worse, according to the use we have made of the opportunities of life. But if any one attempts to define this future state and say we shall have spiritual bodies, live in the skies, sing psalms, and wave palm-branches, we say at once "this is partly unknowable and partly known to be impossible."

These abstract speculations, however, are only adapted for a few of the highest thinkers. That which has given the philosophy of Spencer a wide influence is the manner in which he applies these general principles to the subjects which more immediately concern the mass of thinking minds, such as history, politics, and the problems of social life. What Darwin shows in animal life and the origin of species, Spencer traces in the rise and fall of empires, the growth and decline of religions, the increasing complexity of social relations, the conflicting forces of evolution and dissolution at work around us in our every-day life, in the relations of science and theology, capital and labour, state socialism and *laissez-faire*. For instance, the decline of the Roman Empire and its overthrow by the barbarians is analogous to the decay of a planet from loss of internal heat and its dissipation into matter capable of fresh evolution, by the shock of a comet. The ever-increasing gulf between wealth and poverty, science and superstition, resembles the process by which the one-toed horse became gradually differentiated more and more from the common five-toed type of its remote ancestor.

These speculations of Spencer, pursued with vast acuteness and research through all branches of social science, though they have not founded a new religion or established a new sect, have undoubtedly exercised a great influence on modern thought, especially among the rising generation.

Another "ism" which, although it has exercised a much narrower influence than the philosophy of Spencer, has founded a sect and put forward more definite claims to give the world a new religion, is that which is known as "Positivism," or "Comtism," from the name of its founder, Auguste Comte. It is not easy to understand, but its essence seems to be this

Admitting that science has killed theology, and that the old forms of supernatural religion, inevitable in the childhood of the world, have become incredible, Comte cast about for some idea which should be at the same time "positive," or based on ascertained fact, and fervid enough to satisfy the cravings of religious sentiment. He thought he found it in "Humanity;" that is, in love and veneration for the abstract idea of the human race, taken collectively, and considered in its past, present, and future relations. As patriotism, a very ardent feeling, is the love of a limited section of the human race; and as it has been gradually enlarged from the limits of a tribe to those of a city, and from those of a city to those of a country or nationality, he conceived that it might be still further enlarged so as to embrace all mankind. So far it may be admitted that there is a germ of truth in Comte's idea, and that elevated minds may enlarge their view beyond the narrow bounds of a particular country at a particular period, and may derive fresh incentives to action, and fresh subjects for ennobling thought, from a contemplation of the past progress, present condition, and future possibilities of the collective human race. But there is a homely proverb that "charity begins at home," and as we widen the sphere of patriotism or philanthropy we are very apt to diminish their intensity and find them evaporate in a mist of high-sounding phrases. The "friend of man" is very apt to be the friend of no one man in particular, and to make universal philanthropy an excuse for neglecting individual charity.

Apart, however, from this objection, and granting that with increased intercourse and increased culture

"Humanity" might become a more practical idea, we should be still a long way from making it the basis of a new religion. It is here that Comte has laid himself open to the scoffs of unbelievers, who have gone so far as to call his religion "Catholicism without Christianity," and himself a "grotesque old Frenchman." With the narrow systematising logic so characteristic of the French intellect he has worked out a complete scheme of ritual, hierarchy, and all the apparatus of an old religion. A supreme pontiff at its head, associated with a supreme priestess to represent the female element; for saints the distinguished men who have advanced the different branches of human art and science; for days of worship, fête days of these saints and meetings of believers to commemorate their merits.

All this savours too much of the "Goddess of Liberty," and the theo-philanthropy of the French Revolution, when the disciples of Rousseau cut off heads in the name of universal benevolence, to find much acceptance in a sceptical age and among a practical people. Robuster intellects, like George Eliot, even where they incline to accept Humanity as an ennobling idea, and to recognise Comte as an original thinker, reject all the constructive part of his new religion as unworthy of notice; while to the mass of mankind the whole thing appears utterly unreal and incomprehensible.

One more "ism"—Pessimism, the gospel of feebleness and failure—has had a considerable effect on the Continent, though little in this country. It is based on the fact that, in accordance with the universal law of polarity, progress is not an unmixed good, but develops a corresponding negative of failure. In simple forms of society the distinctions between wealth and poverty, capital and labour, culture and ignorance, are not so sharply defined, and the lot of those who fail in the battle of life is not so hard as when men are congregated in crowded cities, exposed to temptations, and tantalised by the sight of wealth and luxury before their eyes and yet beyond their reach. A mass of misery and discontent is thus created, which in lower natures translates itself into anarchism and fanatical hatred of all above them, while in higher ones it takes the form of theories for the regeneration of the world by levelling everything that exists, and building anew on fresh foundations. Still higher minds see the futility of these theories, and take refuge in a philosophy which pronounces the world a mistake, life an evil, and the only possible solution to be, to put an end to what is radically bad by an act of universal suicide. This is in substance the philosophy of Schopenhauer and the school of Continental Pessimists. It has considerable analogy with that of Buddhism, which considers all personal existence to be a painful dream or illusion, and places supreme happiness in Nirvana, or escape from it by annihilation of individuality.

To understand how such a doctrine can have found acceptance, we must remember that the tendency of modern civilisation is to throw more and more work on the brain and nervous system and less on other organs. This of itself tends to produce more ill-health both of mind and body, especially of those digestive organs upon which the sensation of health and well-being so mainly depends. A dyspeptic man is of necessity an unhappy and desponding man. Moreover, in ruder states of society such weaklings were got rid of by the summary process of being killed off, while with the more humane and refined arrangements of modern times they live on and "weary deaf heaven with their fruitless cries."

It is among such men, with cultivated intellects, sensitive nerves, and bad digestion, that we find the prophets and disciples of the gospel of Pessimism. They feel, and feel truly, that as far as they are concerned life is an evil, the pains of which far outweigh its pleasures, and having lost the mediæval faith in a future life where the balance will be redressed, they see no remedy for the miseries of the world but that of ceasing to be, or annihilation.

This affords another illustration of the extent to which religions and philosophies are, like the spectre of the Brocken, reflections of our own selves on dissolving mists, clothed with our own clothes and repeating our own gestures. To a healthy man or to a strong man the pessimist view of the universe is simply impossible. If he has experienced a fair average of happiness and success in life, he instinctively rejects a creed which tells him that there are no lights as well as shadows. If he has a mind of average strength he feels that suffering is a thing to be avoided prudently, borne stoically, or grappled with courageously, and not to be run away from by moral or physical suicide.

Accordingly Pessimism is not a creed which is ever likely to exert much influence on the strong, practical Anglo-Saxon race, and we can only discern some faint traces of it in the tendency of certain very limited cliques of so-called Æstheticism to admire morbid and self-conscious ideals, both in poetry and painting.

It is a very curious and remarkable fact, that while so many highly intellectual attempts have been made in vain in modern times to found new sects and religions, the only one which has had any real success is that which is based on the most gross and vulgar imposture —Mormonism. Mormonism is a fact which, without the vestige of a reasonable argument to show for itself, originating in the vulgar ravings and forgeries of a vulgar Yankee, violating the first instincts of the family and of society by polygamy, still exists in spite of persecutions, and to a certain extent progresses and flourishes. The reason seems to be that instead of being a theory in the air or over the heads of the masses, it is, with all its faults, a practical system in contact with the actual realities of life. Its success is mainly owing to its being an organised system of emigration, and a faith which places its Paradise here on earth and not in the skies. A poor ignorant labourer in Wales or Norway, who becomes a convert to Mormonism, is taken in hand at once, forwarded to his destination, and when he arrives there looked after and put in a way of earning an honest livelihood and probably becoming a landed proprietor. The ideal set before him is not a very high one, that of becoming a sober, industrious, respectable, narrow-minded citizen of the State of Utah, and a creditable member of the community of Latter Day Saints. But to a poor labourer from the slums of Liverpool, to lead such a life, in the pure mountain air in the valley of the Salt Lake, and see his flocks and herds increasing and his family growing up, without care for the future, is indeed the realisation of an earthly

Paradise. The moral to draw from this is, that the success of a religion, under the conditions of modern society, does not depend so much on its theory as on the way in which it takes hold of the practical problems of life and shows an aptitude for grappling with them.

Another wide-spread modern delusion, that of Spiritualism, is akin to Mormonism, as showing how little reason has to do with the beliefs which are most readily propagated among large classes of the community. Nothing but the most morbid appetite for the supernatural, combined with the most absolute ignorance of the laws of evidence, could induce sane people to believe that, if a corner of that mysterious and awful veil were lifted which separates the living from the dead, we shall discover what?—spirits whose vocation it is to turn tables and talk twaddle.

In vain medium after medium is detected, and the machinery by which ghosts are manufactured exposed in police-courts; in vain the manifestations of the so-called spirits are repeated by professional conjurers like Maskelyne and Cooke, who disclaim any assistance from the unseen world. People are still found to believe the unbelievable because it gratifies their taste for the marvellous, and enables them to fancy themselves the favoured recipients of supernatural communications.

If Spiritualism has found a certain amount of acceptance from men of a very different order, who, like Crookes and Wallace, understand what scientific evidence really is, it is because the phenomena associated with it, such as mesmerism and clairvoyance, really have a certain basis of fact, and open up interesting fields for scientific investigation. The working of the brain and nerves in certain abnormal conditions, and the physical effects of imagination, are subjects imperfectly understood, but which well deserve accurate inquiry.

Take, for instance, dreams, which afford the first certain starting-point towards the theory of visions and apparitions. It is as certain that we dream as that we sleep, and that in our sleeping state we often live a sort of second life, which is different from our ordinary waking life. Dreams seem to be made up of impressions which have been photographed on the brain in its waking state, and which are revived and worked up into new combinations and imaginary scenes, when consciousness is suspended. Vivid impressions are thus often worked up into a succession of dreams so vivid as to be scarcely distinguishable from reality. It happened to me, about the middle period of my life, to be sent, almost at a day's notice, to India, where for more than two years I had a period of intensely hard work and great responsibility, as Finance Minister. This naturally left a number of strong impressions on my brain, which for years afterwards kept reviving in a series of connected dreams, in which I fancied myself back in India. I had thus a dream life as well as a real life of Indian experiences, and the former was so vivid that, if I were writing reminiscences, I should sometimes find it difficult to distinguish between the two.

This enables me to realise how dreams may readily pass into visions. If I had dozed off in an arm-chair after dinner, and fallen into one of my Indian dreams, I might have seen Lord Canning, who had been dead for years, walk into the room as distinctly as if he had been present in person. In a less critical age, and with a less sceptical turn of mind, I might readily have been convinced that I had seen his ghost.

There can be no doubt that in this way dreams must often, in pre-scientific ages, have originated a *bonâ fide* belief in spirits. Herbert Spencer traces to this cause the origin of all religious belief. Perhaps this may be carrying it too far, but doubtless it was one of the main causes, especially of that portion of religion which took the form of offerings to the dead and ancestor-worship.

But a still further step may be taken from the ordinary dream to the waking dream or vision. It is a well-established fact that under peculiar and rare circumstances the brain may dream, that is, revive impressions where there is no corresponding reality, without losing its consciousness. There was a celebrated case of a Berlin bookseller in the last century, who, having fallen into bad health, lived for more than a year in the company of ghosts—that is, he constantly saw men and women, with every appearance of being alive, enter the room and come and go as if they had been ordinary visitors. Being a man of a scientific turn of mind he never supposed these were really ghosts, but reasoned on them and recorded his experiences. Instead of sending for a priest and resorting to exorcisms he called in a physician and took a course of medicine, with the result that after a considerable time the ghostly visitors gradually became dim and finally disappeared.

Numerous other cases are recorded in which there is no doubt that visions have been seen, especially under the influence of religious excitement, and a large number of so-called miraculous appearances and ghost stories are probably owing to this cause rather than to conscious imposture.

When we consider the enormous number of dreams, and probably considerable number of visions, which occur, instead of being surprised at occasional coincidences, the wonder rather is that they are not more frequent. If only one per cent. of the 30,000,000 inhabitants of the British Isles dream every night, that would give 109,500,000 dreams per annum, a large proportion of which are made up of vivid impressions of actual persons and events. It is impossible that some of the combinations of these impressions should not form pictures which are subsequently realised, and we may be sure that the successes only will be noted, and the failures forgotten. It is strange, therefore, that the researches of the Psychical Society should not have brought to light more instances of death-warnings and other remarkable coincidences. To take the vulgar instance of horse-racing. A number of minds are greatly exercised over the problem of picking out winners, and doubtless a vast number of dreams show colours flashing past winning-posts, and numbers hoisted on the telegraph board. And yet I only remember two tolerably well-authenticated instances in the last half-century, in which any one is said to have backed a winner on the faith of a dream. The only positive result of dreams and visions is that they frequently occur under circumstances where they are almost certain to be mistaken, by unscientific persons in unscientific ages, for actual supernatural appearances.

Another field for inquiry is opened out by the

effects which are undoubtedly produced under certain abnormal conditions of the brain and nervous system, as in epilepsy, somnambulism, and mesmerism.

In the simplest case, that of epilepsy, the effect is mainly shown by a more intense action of nerve-currents, causing convulsive motions and an unnatural increase of muscular strength and rigidity, so that two strong men may be scarcely able to hold one weak woman. In somnambulism, the effects are more complex. The reception of outward impressions seems to be limited, so that the whole consciousness and vital energy are concentrated on particular actions, which are thus performed safely, while in the ordinary waking state they would be impossible. Thus a somnambulist walks securely along a plank spanning an abyss, because the impressions of surrounding space do not reach the brain and confuse it with a sense of danger. In this state also past impressions photographed on the brain, which in the ordinary waking state are obscured by other impressions, seem to come out occasionally as in dreams, enabling the somnambulist to do and remember things which would otherwise be beyond his faculties.

Mesmerism is closely akin to somnambulism. Apart from delusion and charlatanism the fact seems to be established that it is possible, by artificial means, to induce a state resembling somnambulism in persons of a peculiar nervous temperament. As regards the means, the essential point seems to be to throw the brain into this abnormal state partly by keeping an unnatural strain on the attention, and partly by acting on it through the imagination. The experiments of Dr. Braid showed that the mesmeric sleep could be induced just as well by keeping the eye strained on a black wafer stuck on a white wall, as by the manipulations of an operator. This experiment disposes of a great deal of mysterious nonsense about magnetic fluids, overpowering wills, and other supposed attributes of professional mesmerisers, and reduces the question to the plain matter-of-fact level of the relations between the brain, will, imagination, and nervous system, which exist in natural and in artificial somnambulism. These are undoubtedly very curious, and open up a wide field for physiological and mental research. As far as I have seen or read, they seem to turn mainly on the reflex effects of an excited imagination on other organs and faculties. I do not believe that any one could be mesmerised who was absolutely ignorant of the subject and unconscious that any one was operating. On the other hand, any one who had frequently been mesmerised would fall into the sleep if led to believe that an operator was at work when there was really not one. And the peculiar effects shown in the mesmeric state are attributable mainly, if not entirely, to the imagination acting with morbid activity on the slightest hint or suggestion of what is expected. Thus the will disappears in the more powerful suggestion of the imagination that the patient has to obey the will of the operator, or do certain things which are in the programme. I can readily believe also that in this state the imagination can perform feats which would be impossible to it in a natural state when it is kept in check by other faculties, and that a good deal of what is called clairvoyance may be explained by the way in which the slightest hint from expression, involuntary muscular motion, or otherwise, is taken advantage of as a substitute for the ordinary modes of communication. Such a faculty may also doubtless be cultivated by practice, and thus explain many of the phenomena of what are called spiritual communications and thought-reading. But that impressions can be made on the brain, or that one mind can communicate with another, without some physical means of connection between object and subject, is absolutely unproved and remains altogether incredible.

Among the great writers who, without attempting to found sects, have profoundly influenced modern thought, Carlyle undoubtedly occupies the foremost place. With all his extravagances and eccentricities, he was essentially a Hebrew prophet in modern guise, preaching a true gospel—that of sincerity. To stand on fact and despise shams, to make one's life accord with the "eternal veracities," to strip off outward trappings and look at the ideas they clothe, to worship truth and abhor falsehood; these are the principles which Carlyle is never tired of enforcing in his vivid and picturesque language. The dignity of all faithful work, and the hollowness of mere show and pretence, is another theme on which he delights to dwell; and the maxim, "Do the nearest duty that lies to your hand and already the next duty will have become plainer," is his favourite rule for practical conduct. He insists much on "hero-worship," and pushes his conclusions to an extreme extent, dividing mankind too absolutely into two classes; on the one hand the heroes who discern facts and the followers who loyally obey them; on the other, the great mass of foolish and chattering humanity. Human nature is not really all black or all white, but shaded off by innumerable half-tints and blended gradations. Nevertheless "hero-worship" contains a great truth, that loyal reverence for what we feel to be above us does not lower a man but elevates him; and that those really degrade themselves who have no respect for higher things, and try to drag everything down to their own level.

In insisting on looking through phrases to facts Carlyle touches one of the great dangers of the present day. The spread of education has given an extension to the influence of words which threatens to become excessive. People read until they have no time to think, and find it easier to borrow the thoughts of others. And a large and ever-increasing portion of the community have learnt, in Yankee phrase, to "orate," and use the new-found faculty incessantly and remorselessly. I do not refer so much to the obstruction of the Parliamentary machine by floods of talk, for that is an evil which will work its own cure, but to the undue influence which oratory tends to acquire in all constitutional countries, where the ultimate power is vested in what is essentially a debating society. A great orator is inevitably a great power in the State, but it does not necessarily follow that he is a great statesman.

The qualities which make an orator depend to a great extent on gifts of nature, such as a good voice and presence, and still more on the gift of a fervid temperament, which moves and convinces others because the speaker is himself moved and convinced. These may or may not coincide with the gifts of a great statesman, ripe experience, clear judgment, and calm courage. When they do coincide the State will be well ruled; when they do not, the statesman will lack motive power, and the orator will lack statesmanship; so that between the two the affairs of the nation will be apt to be mismanaged. Still, on the whole we must accept the inevitable, and trust that the public opinion which is formed by many speeches and many articles will give better average results than by attempting to find a Hero who might just as readily turn out to be a Cleon as a Pericles. But the influence of Carlyle's teaching will always remain useful as a corrective, and as a warning to public opinion to measure public men by their solid qualities rather than by their oratorical talent.

The influence of Carlyle has been great on all the foremost minds of his generation, and may be distinctly traced in their writings. If Tennyson makes his Guinevere say:

> Oh God! what might I have made of Thy fair earth
> Had I but loved Thy highest creature in it!
> We needs must love the highest when we see it.

This is genuine Carlylese condensed into noble poetry. The whole literature of fiction has been transformed. The fashionable novel, with its dandified coxcomb heroes and simpering fine lady heroines, has been superseded by works like those of Dickens, Thackeray, Trollope, and George Eliot, which satirise folly and pretension however highly placed, and aim at honest, earnest, simple and sincere ideals of true men and women. The whole tone of society has become more manly, and no one now thinks of acquiring fame by wearing a pea-green coat or getting a voucher for Almack's. Artificial distinctions have to a great extent disappeared, the sons of dukes think it no disgrace to earn an honest livelihood as stockbrokers, and self-made men are received on an equal footing everywhere if they have the essential qualities of gentlemen. There is vastly more real equality and real fraternity among men, and every one recognises, in theory at any rate, the dignity of honest labour, whether of the hand or head. The certain survival also, in the long run, of truth over falsehood, or in other words of the fittest, as being most in accordance with the laws of the universe, is universally recognised as a law in the moral as well as in the material world.

For these results, which have now become almost commonplaces, those who derived them in their youth direct from works like "Sartor Resartus" can best judge to what an extent modern thought has been indebted to Carlyle.

CHAPTER VIII.

MIRACLES.

Origin of Belief in the Supernatural—Thunder—Belief in Miracles formerly Universal—St. Paul's Testimony—Now Incredible—Christian Miracles—Apparent Miracles—Real Miracles—Absurd Miracles—Worthy Miracles—The Resurrection and Ascension—Nature of Evidence required—Inspiration—Prophecy—Direct Evidence—St. Paul—The Gospels—What is Known of Them—The Synoptic Gospels—Resemblances and Differences—Their Origin—Papias—Gospel of St. John—Evidence rests on Matthew, Mark, and Luke—What each states—Compared with one another and with St. John—Hopelessly Contradictory—Miracle of the Ascension—Silence of Mark—Probable Early Date of Gospels—But not in their Present Form.

WHEN men began to reason on the phenomena of the world around them, it was inevitable that they should begin by referring all striking occurrences to supernatural causes. Just as they measured space by feet and inches, and time by days and years, they referred unusual events to personal agencies. They knew by experience that certain effects were produced by their own wills, muscular energies, and passions; and when they saw effects which seemed to be of a like nature, they inferred that they must have been produced by like causes.

To take the familiar instance of thunder. The

first savage who thought about it must have said: "The sound is very like the roar with which I spring on a wild beast or an enemy; the flash of lightning is very like the flash of the arrow or javelin with which I strike him; the effect is often the same, that he is killed. Surely there must be some one in the clouds, very strong, very angry, very able to do me harm, unless I can propitiate him by prayers or offerings." But after long centuries, science steps in. An elderly gentleman at Philadelphia, Benjamin Franklin by name, sends up a silk kite during a thunder-storm, and behold! the lightning is drawn down from the skies, tamed, and made to emit harmless sparks, or to follow the course of a conducting wire, at our will and pleasure. There is no more room left for the supernatural in the fiercest tropical thunder-storm than there is in turning the handle of an electrical machine, or sending in a tender to light the streets of London by electric light. And the result is absolutely certain. In the contest between the natural and the supernatural, the latter has not only been repulsed but annihilated. The most orthodox believer in miracles, if his faith were brought to the practical test of backing his opinions by his money, would rather insure a gin-palace or gambling saloon protected by a lightning-conductor than a chapel protected by the prayers of a pious preacher.

This instance of thunder is a type of the revolution of thought which has been brought about by modern science in the whole manner of viewing the phenomena of the surrounding universe. Former ages saw miracles everywhere, the age in which we live sees them nowhere, except possibly in the single instance of the miracles recorded in the Bible. In the annals of grave Roman historians,

> In every page *locutus bos*.

Not a Cæsar or a Consul died, without an ox speaking, or a flaming sword in the skies predicting portents. If the moon happened to pass between the sun and the earth the dim eclipse

> With fear of change perplexes monarchs.

If the winds blow it is because Æolus releases them from the cave; if the rains fall it is because Jupiter opens the windows of heaven, or Indra causes the cloud-cows to drop their milk on the parched earth. Perhaps no better proof can be afforded of the universal belief that miracles were considered matters of every-day occurrence than is given by the passage in St. Paul's Epistle to the Corinthians, in which he enumerates the principal Christian gifts, and assigns, as it were, their comparative order and the number of marks that should be given to each in a competitive examination.

The power of "working miracles" comes low in the list. "First apostles, secondarily prophets, thirdly teachers, after that miracles, then gifts of healings, helps, governments, diversities of tongues." And he goes on to say, in words that come home to every heart in all centuries, that all those things are worthless as compared with that true Christian charity which "suffereth long, and is kind; envieth not; vaunteth not itself, is not puffed up, doth not behave itself unseemly, seeketh not her own, is not easily provoked, thinketh no evil; rejoiceth not in iniquity, but rejoiceth in the truth; beareth all things, believeth all things, hopeth all things, endureth all things."

This is in the true spirit of modern thought, which, when the externals of religion fail, strives to look below them at its essence, and to retain what is eternally true and beautiful as the ideal of a spiritual and the guide of a practical life, while rejecting all the outward apparatus of metaphysical creeds and incredible miracles, which had only a temporary value, and can no longer be believed without shutting one's eyes to facts and becoming guilty of conscious or unconscious insincerity.

But to return to miracles. Almost the entire world of the supernatural fades away of itself with an extension of our knowledge of the laws of Nature, as surely as the mists melt from the valley before the rays of the morning sun. We have seen how, throughout the wide domains of space, time, and matter, law, uniform, universal, and inexorable, reigns supreme; and there is absolutely no room for the interference of any outside personal agency to suspend its operations. The last remnant of supernaturalism, therefore, apart from the Christian miracles which we shall presently consider, has shrunk into that doubtful and shady border-land of ghosts, spiritualism and mesmerism, where vision and fact, and partly real partly imaginary effects of abnormal nervous conditions, are mixed up in a nebulous haze with a large dose of imposture and credulity.

Even this region is being contracted every day by every fresh revelation in a police-court, and every fresh discovery of the laws which really regulate the transmission of nervous energy to and from the brain, in the abnormal state which constitutes epilepsy and somnambulism, and enables an excited imagination to produce physical effects, such as those of drastic drugs on a patient who has actually taken nothing but pills of harmless paste.

The question of Christian miracles, however, rests on a different and more serious ground. They have been accepted for ages as the foundation and proof of a religion which has been for nineteen centuries that of the highest civilisation and purest morality, and for this reason alone they deserve the most reverent treatment and the most careful consideration.

Of a large class of these miracles it may be said that there is no reason to doubt them, but none to consider them as violations of law, or anything but the expression, in the language of the time, of natural effects and natural causes. When a large class of maladies were universally attributed to the agency of evil spirits which had taken possession of the patient's body, it was inevitable that many cures would be effected, and that these cures would be set down as the casting out of devils. In many cases also a strong impulse communicated to the brain may send a current along a nerve which may temporarily, or even permanently, restore motion to a paralysed limb, or give fresh vitality to a paralysed nerve. Thus, the lame may walk, the dumb speak, and the blind see, with no more occasion to invoke supernatural agency than if the same effects had been produced by a current of electricity from a voltaic battery. There is no reason to doubt that miracles of this sort have been frequently wrought by saints and relics, and that even at the present day they may possibly be wrought at Lourdes and other shrines of Catholic faith. Only at the present day we scrutinise the evidence and count the failures, and admit nothing to be supernatural which can be explained as within a fair average result of exceptional cases under the operation of natural laws. In like manner we set down all visions or apparitions as having no objective reality if they can be explained by the known laws of dreams or other vivid revivals of impressions, on the brain of the person who perceives them.

There remains the class of really supernatural miracles, or miracles which could by no possibility have occurred as they are described, unless some outward agency had suspended or reversed the laws of Nature. As regards such miracles, a knowledge of these laws enormously increases the difficulty in believing in them as actual facts. Take for instance the conversion of water into wine. When nothing was known of the constitution of water or of wine, except that they were both fluids, it was comparatively easy to accept the statement that such a conversion really took place. But now we know that water consists of oxygen and hydrogen combined in a certain simple proportion, and of these and nothing else; while wine contains in addition nitrogen, carbon, and other elements combined in very complicated proportions. If the water was not really changed into wine, but only seemed to be so, it was a mere juggling trick, such as the Wizard of the North can show us any day for a shilling. But if it was really changed, something must have been created out of nothing to supply the elements which were not in the original water and were not put into it from without.

Again, those who have followed the question of spontaneous generation, and witnessed the failure of the ablest chemists to produce the lowest forms of protoplasmic life from inorganic elements, will hardly believe that such a highly organised form of life as a serpent could have been really produced from a wooden rod. And this, be it observed, not only by Moses the prophet of God, but by the jugglers who amused the court of Pharaoh by their conjuring tricks; and for an object of no greater moment than to persuade a king to allow some of his subjects to emigrate, which object, moreover, notwithstanding the miracle, entirely failed, as the king simply "hardened his heart" and persisted in his refusal.

But passing from this class of grotesque and incredible miracles, let us examine those which may be called worthy miracles; that is, miracles disfigured by no absurd details, and wrought for objects of sufficient importance to justify supernatural interference, if ever such interference were to take place. At the head of such miracles must undoubtedly be placed those of the Saviour's resurrection. The appearances to the Apostles, and above all the bodily Ascension to heaven in the presence of more than 500 witnesses, were a fitting termination to the drama of His life and sufferings, and afforded a conclusive test of the fact which was the foundation-stone of the new religion.

"If Christ be not risen, then is our preaching vain," says St. Paul; and he proceeds to argue that the whole question of the reality of a future life hinges on the fact that Christ really rose from the dead. His theory is that death came into the world by the sin of the first man, Adam, and has been destroyed and swallowed up in immortality by the victory of the second man, Christ. This theory has, from that day to this, been the key-stone of Christian theology.

There can be no doubt, therefore, that if any miracle is true this must be the one, and, on the other hand, if this miracle cannot be established by sufficient proof, it is idle to discuss the evidence for other miracles. In order to go to the root of the matter therefore, it is necessary to consider, in a calm and judicial spirit, the evidence upon which this miracle of the Resurrection really rests.

In the first place we must consider what sort of evidence is required to prove a miracle. Clearly it must be evidence of the most cogent and unimpeachable character, far more conclusive than would be sufficient to establish an ordinary occurrence. The discoveries of modern science have shown beyond the possibility of doubt that the miracles which former ages fancied they saw around them every day had no real existence, and that, except possibly in the solitary instance of the Christian miracles, there has been no supernatural interference with the laws of Nature throughout the enormous ranges of space, time, and matter. It may be going too far to say with Hume that no amount of evidence can prove a miracle, since it must always remain more probable that human testimony should be false than that the laws of Nature should have been violated. But it is not going too far to say that the evidence to establish such a violation must be altogether overwhelming and open to no other possible construction.

Take the case of the allegation that a man who had really died rose in the body from the grave, ate, drank, and held intercourse with living persons. There are some 1,200 millions of human beings living in the world, and somewhat more than three generations in each century, that is, there are some 3,600 millions of deaths per century, and this has been going on for some forty or fifty centuries, or longer. It is certain, therefore, that at least 150,000 millions of deaths must have taken place, and a large proportion of these under circumstances involving the most heart-rending separations, and the most intense longing on the part of the dying to give, and of the living to receive, some token of affection from beyond the grave. And yet no such token has ever been given, and the veil which separates the dead from the living has never been lifted, except possibly in one case out of this 150,000,000,000. Surely it must require very different evidence to establish the reality of such an exception, from that which would be sufficient to prove the signature to a will or the date of a battle.

But just when the new views opened up by modern science made it more difficult to believe in miracles, and more exacting in the demand for stronger evidence to support them, the old evidence became greatly weakened. The main evidence which satisfied our forefathers was that the Bible was inspired, and that it asserted the reality of the miracles. This, when critically examined, was really no evidence at all, for how did we know that the Bible was inspired? Because it was proved to be so by miracles. The argument was therefore in a circle, and resembled that of the Hindoo mythology, which rested the earth on an elephant and the elephant on a tortoise. But what did the tortoise rest on?

To examine the matter more closely, what is the meaning of inspiration? It means that a certain book was not written, as all other books in the world have been written, by writers who were fallible, and whose statements and opinions, however admirable in the main and made in perfect good faith, inevitably reflected the views of the age in which they lived and contained matters which subsequent ages found to be obsolete or erroneous, but that this particular book was miraculously dictated by an infallible God, and therefore absolutely and for all time true. But, as a chain cannot be stronger than its weakest link, if any one of these statements were proved not to be true, the theory of inspiration failed, and human reason was called on to decide by the ordinary methods, whether any, and if any, what parts of the Volume were inspired and what uninspired.

Now it is absolutely certain that portions of the Bible, and those important portions relating to the creation of the world and of man, are not true, and therefore not inspired. It is certain that the sun, moon, stars, and earth, were not created as the author of Genesis supposed them to have been created, and that the first man, whose Palæolithic implements are found in caves and river gravels of immense antiquity, was a very different being from the Adam who was created in God's likeness and placed in the Garden of Eden. It is certain that no universal deluge ever took place since man existed, and that the animal life existing in the world, and shown by fossil remains to have existed for untold ages, could by no possibility have originated from pairs of animals living together for forty days in the ark, and radiating from a mountain in Armenia.

Another test of inspiration is afforded by the presence of contradictions. If one writer says that certain events occurred in Galilee, while another says that they took place at Jerusalem, they cannot both be inspired. They may be both reminiscences of real events, but they are obviously imperfect and not inspired reminiscences, and require to be tested by the same process of reasoning as we should apply in endeavouring to unravel the truth from the confused and contradictory evidence of conflicting historians.

Inspiration is clearly as much a miracle as any of the miracles which it relates, and there is only one way conceivable by which it could be proved, so as to afford a solid basis for faith and give additional evidence in support of the supernatural occurrences said to have taken place; that would be if it carried with it internal evidence of its truth. Such evidence might be afforded in one way, and in one only—by prophecy. If any volume written many centuries ago contained a clear, definite, and distinct prophecy of future events, which the writer could by no possibility have known or conjectured, such a prophecy must have been dictated by some agency different from anything known in the ordinary course of nature; and future ages, seeing the fulfilment of the prophecy, could scarcely doubt that the volume which contained it was inspired. But such a prophecy must be quite definite, so that there could be no doubt as to whether it had been fulfilled or not, and must not consist of vague and mystic utterances, in which future believers might find meanings, probably never thought of by the prophets themselves, confirming the faith which, from other considerations, they thought it a sin to disbelieve. Nor must it consist of passionate aspirations for deliverance, and predictions of the downfall of cruel conquerors, wrung from the hearts of an oppressed people in times of imminent danger and crushing despair; because such predictions have been partly verified and partly transformed in future ages, so as to receive a new and spiritual significance.

There is one prophecy which affords a test by which to judge of the value of all others as a proof of inspiration, for it is perfectly distinct and definite, and comes from the highest authority—that of the approaching end of the world contained in the New Testament.

St. Matthew reports Jesus to have said:

"For the Son of man shall come in the glory of his Father with his angels; and then he shall reward every man according to his works.

"Verily I say unto you, There be some standing here, which shall not taste of death, till they see the Son of man coming in his kingdom."

It is certain that all standing there did taste death without seeing the Son of Man coming with His angels. The conclusion is irresistible, that either Jesus was mistaken in speaking these words, or else Matthew was mistaken in supposing that He spoke them.

St. Paul predicts the same event in still more definite terms. He says:

"For this we say unto you by the word of the Lord, that we which are alive and remain unto the coming of the Lord shall not prevent them which are asleep.

"For the Lord himself shall descend from heaven with a shout, with the voice of the archangel, and with the trump of God: and the dead in Christ shall rise first:

"Then we which are alive and remain shall be caught up together with them in the clouds, to meet the Lord in the air."

Here is the most distinct prediction possible, both of the event which was to happen and of the limit of time within which it was to take place; and, to give it additional force, it is specially declared to be an inspired prophecy uttered as "the word of God."

The time is distinctly stated to be in the lifetime of some of the existing generation, including Paul himself, who is to be one of the "we which are alive," who are not to "prevent," or gain any precedence over, those who have "fallen asleep," or died, in the interval before Christ's coming. By no possibility can this be construed to mean a coming at some indefinite future time, long after all those had died who were to remain and be caught up alive into the clouds. St. Paul doubtless meant what he said, and firmly believed that he was uttering an inspired prophecy which would certainly be fulfilled. But it is certain that it was not fulfilled. Paul and all Paul's contemporaries have been dead for 1,800 years, and the shout, the voice of the Archangel, and the trump of God, have never been heard. What is this but an absolutely irresistible demonstration that prophecy not only fails to prove inspiration, but, on the contrary, by its failure disproves it, and shows that St. Matthew and St. Paul were as liable to make mistakes as any of the hundreds of religious writers who, in later times, have prophesied the approaching end of the world or advent of the millennium.

The evidence for miracles, therefore, must be taken

on its own merits, without aid from any preconceived theory that it is sinful to scrutinise it because the books in which it is contained are inspired. Applying to it impartially the ordinary rules of evidence, let us see what it amounts to, for that which is really the test case of all other miracles, that of the Resurrection.

The witnesses are St. Paul and the authors of the four Gospels according to St. Matthew, St. Mark, St. Luke, and St. John. Of these, St. Paul is in some respects the best. When a witness is called into court to give evidence, the first question asked is, "Who are you? Give your name and description." St. Paul alone gives a clear answer to this question. There is no doubt that he was an historical personage, who lived at the time and in the manner described in the Acts of the Apostles, and that the Epistle to the Corinthians is a genuine letter written by him. In this Epistle he says:

"For I delivered unto you first of all that which I also received, how that Christ died for our sins according to the scriptures;

"And that he was buried, and that he rose again the third day according to the scriptures:

"And that he was seen of Cephas, then of the twelve:

"After that, he was seen of above five hundred brethren at once; of whom the greater part remain unto this present, but some are fallen asleep.

"After that, he was seen of James; then of all the apostles.

"And last of all he was seen of me also, as of one born out of due time."

This is undoubtedly very distinct evidence that

the appearances described by St. Paul were currently believed in the circle of early Christians at Jerusalem within twenty years of their alleged occurrence.

This is strong testimony, but it is weakened by several considerations. In the first place, we know that Paul's frame of mind in regard to miracles was such as to make it certain that he would take them for granted, and not attempt to examine critically the evidence on which they were founded, and this was doubtless the frame of mind of those from whom he received the accounts. Again, he places all the appearances on the same footing as that to himself, which was clearly of the nature of a vision, or strong internal impression, rather than of an objective reality. Upon this vital point, whether the appearances which led to the belief in Christ's resurrection were subjective or objective—that is, were visions or physical realities —Paul's testimony therefore favours the former view, which is quite consistent with the laws of Nature and with experience in other cases.

And finally, St. Paul's account of the appearances is altogether different from those of the other witnesses, viz., the four Evangelists.

When we come to consider the testimony of the four Gospels we are confronted by a first difficulty: Who and what are the witnesses? What is really known of them is this: Until the middle of the second century they are never quoted, and were apparently unknown. Somewhere about 150 A.D., for the exact date is hotly disputed, we find the first quotations from them, and from that time forwards the quotations become more frequent and their authority increases, until finally they superseded all the other narratives current in the early

Church, such as the "Gospel of the Hebrews," and the "Pastor" of Hermas, and are embodied in the canon of inspired writings of the New Testament. From the earliest time where there is any distinct recognition of them, they appear to have been attributed to the Evangelists whose names they bear, viz., Matthew, Mark, Luke, and John.

When we look to internal evidence to give us some further clue as to their authorship and date, we at once meet with a great difficulty. The three Gospels of St. Matthew, Mark, and Luke, are called "Synoptic," because they give what is substantially the same narrative of the same facts arranged in the same order, and the same sayings and parables giving the same view of the character and teaching of Jesus. In whole passages this resemblance is not merely substantial but literal, so that we cannot suppose it to arise merely from following the same oral tradition, and cannot doubt that the authors must have copied verbatim either from one another or from some common manuscript. But then comes in this perplexing circumstance. After passages of almost literal identity come in statements which are inconsistent with those of the other Gospels and narratives of important events which are either altogether wanting or quite differently described in them.

Thus, in the vital matter of the Resurrection, Matthew says that the disciples were especially commanded to "go into Galilee; there shall you see him," and that they did go accordingly, and there saw Him on a mountain where He had appointed them to meet Him; while Luke distinctly says that "he commanded them that they should not depart from Jerusalem," and describes them as remaining there and witnessing a number of appearances, including the crowning miracle of the Ascension (the same, doubtless, as that which St. Paul describes as having taken place in the presence of more than 500 witnesses), of which Matthew, Mark, and John apparently know nothing. And yet the final injunction of Jesus to preach the gospel in His name to all nations is given in almost the same words in Matthew, Mark, and Luke, showing that they must have had before them some common manuscript describing the course of events after the Crucifixion.

So in minor matters, Mark mentions the cure of one blind man, Bartimæus, who sat by the roadside begging; in Matthew there are two blind men, and yet the dialogue that passed—"What will ye that I shall do unto you?" "Lord, that our eyes may be opened"—is almost word for word the same. It would seem that if they did copy from an original manuscript, they felt themselves free to take any liberties with it they liked, in the way of omission and alteration.

The only light thrown on this perplexing question of the origin of the Gospels is that afforded by the celebrated passage from Papias quoted by Eusebius. Papias was Bishop of Hieropolis, in Asia Minor, and suffered martyrdom, when an aged man, about the year 164. He was therefore brought up in personal contact, not with the Apostles themselves, but with those who, like Polycarp and others, had been their immediate disciples, and had known and conversed with them. In the passage quoted he states his preference for oral tradition over written documents, and his reasons for it. He says: "If I found some one who had followed the first presbyters, I asked him what he had heard from them; what said Andrew or Peter, or Philip, Thomas, James, John, or Matthew; and what said Andrew and John the Presbyter, who were also disciples of the Lord; for I thought I could not derive as much advantage from books as from the living and abiding oral tradition." And he goes on to give his reasons for not attaching more weight to the two written sources of information which were evidently best known and looked upon as of most authority in his time, viz., the Gospels according to St. Matthew and St. Mark. He says that Matthew wrote down in Hebrew the Logia, or principal sayings and discourses of the Lord, "which every one translated as he best could," evidently implying that these numerous translations were, in his opinion, loose, inaccurate, and unreliable. As regards Mark, he says that "Mark, who had not known the Lord personally, and had never heard Him, followed Peter later as his interpreter; and when Peter, in the course of his teaching, mentioned any of the doings or sayings of Christ, took care to note them down exactly, but without any order, and without making a continuous narrative of the discourses of the Lord, which did not enter into the intention of the Apostle. Thus Mark let nothing pass, jotting down a certain number of facts as Peter mentioned them, but having no other care than to omit nothing of what he heard, and to change nothing in it."

This testimony of Papias is very valuable and very

instructive. In the first place, it seems conclusive that the Gospel of St. John was not known to him, and not received in the early Christian Churches of Asia Minor as a work of authority. Had it been so received, Papias must have known of it, brought up as he was at the feet of men who had been John's disciples, and bishop of a Church closely connected with those of which, if there is any faith in tradition, John had been the patriarch and principal founder. And if he had known of such a written Gospel as that of St. John, and believed it to have been really written by the "beloved disciple," the Apostle second only, if second, to St. Peter, it is inconceivable that he should have expressed such an unqualified preference for oral tradition, and made such an almost contemptuous reference to written documents. He must have said: "For, with the exception of the Gospel of the blessed John, I found that little was to be got from books."

It seems clear, therefore, that although the Gospel of St. John may contain genuine reminiscences of an early date, and possibly some which really came from the Apostle himself, the work in its present form could not have been written by him, and must have been compiled at such a late date as to have been unknown in the Christian Churches of the East in the time of Papias.

The same remark applies to the Gospel of St. Luke, of which Papias has equally no knowledge, and which, from internal evidence, appears to be a later edition of the two earlier Gospels, or of the original manuscripts from which they were taken, altered in places to meet objections of a later date, as where the injunction to "go into Galilee; there shall ye see him," is changed into "as he spake unto you when he was yet in Galilee," obviously to reconcile the statement with the subsequent belief that the Ascension took place at Jerusalem.

There remain the two original Gospels according to St. Matthew and St. Mark. Volumes of erudition have been written to try and reconcile them with one another, and with the other two Gospels, and to explain the extraordinary resemblances and no less extraordinary differences. Translations have been heaped on translations, and successive editions and revisions piled on one another until the edifice toppled over by its own weight, but, after all, we have nothing better to rely on than the statement of Papias, which there is no reason to mistrust. The basis of the three Synoptic Gospels was probably a collection of facts and anecdotes written down in Greek by Mark, and of discourses written in Hebrew by Matthew. These have been worked up subsequently, at unknown dates, and by unknown authors, aided possibly by oral traditions, into connected narratives or biographies of the life and teachings of the Founder of the religion.

Possibly, though by no means certainly, we have in the present Gospel according to St. Matthew the nearest approach to the original Logia or doctrinal discourses, and in the present Mark the nearest approach to the original notes recorded by Mark from the dictation of St. Peter.

As regards the Gospel according to St. John, it appears perfectly clear, both from the silence of Papias, the absence of any reference to it by other early Christian Fathers until the end of the second century, and still more from internal evidence, that it could not possibly have been written by the Apostle whose name it bears. John, as we know from St. Paul's Epistles, was one of the pillars of the Christian Church of Jerusalem, whose doctrine was in all respects Hebraic, and who opposed the larger idea that a man could be a Christian without first becoming a Jew.

The writer of the Gospel is not only ignorant of matters which must have been well known to every Jew, but he is positively prejudiced against Judaism, and represents it in an unfavourable light. His narrative of the events of the life of Jesus, including the miracles, is totally different from that of the Synoptics, and his view of His character and report of His speeches wide as the poles asunder. To the Synoptics Jesus is the man-Messiah foretold by the prophets; to the author of John He is the "Logos," the incarnation of a metaphysical attribute of the Deity.

The terse and simple clearness of His sayings recorded by the first, is exchanged in the latter for an involved and cumbrous phraseology reminding one of a Papal Encyclical. The amiability and "sweet reasonableness" of the Jesus of the Synoptics, have become acrimonious unreasonableness and egotistical self-glorification in many of the long harangues which are introduced on the most unlikely occasions in the fourth Gospel.

It is evident, therefore, that this Gospel can afford no aid towards a critical examination of contemporary evidence, and that for this we must look almost entirely to such remains of early records as are preserved in the Gospels of St. Matthew, St. Mark, and St. Luke. With these data, how does the evidence stand as regards the miracles of the Resurrection which are the test cases of all alleged miracles?

It is important to observe that the oldest manuscripts of the Gospel of St. Mark stop at the 8th verse of the last chapter, and that the subsequent verses, 9—20, have every appearance of being a later addition made to reconcile this Gospel better with the prevailing belief and with the other Gospels. Commentators discover a difference in the style and language, and the appearances are described in vague and general language, very different from the distinct details given of them in the other Gospels, and inconsistent with the formal statement twice repeated in the genuine Mark that they were to take place in Galilee. Moreover, if these verses were really in the original Gospel, it is inconceivable how they should have dropped out in the oldest manuscripts, while it is perfectly conceivable how they should have been added at a later period, when the Fathers of the Church began to occupy themselves with the task of harmonising the different Gospels.

But if the genuine Mark really terminated with the 8th verse, not only is there no confirmation of the four miraculous appearances, including the Ascension, recorded by St. Paul as being currently believed by the early Christians within twenty years of their occurrence, but there is positively no mention of any appearance at all. A young man, clothed in white, tells three women who went to the tomb that Jesus is risen, and that they were to tell His disciples and Peter that they would see Him in Galilee; an injunction which was not carried out, for the women "were afraid, neither said they anything to any man."

In St. Matthew the young man has become an angel, and as the women return from the tomb Jesus met them and said, "All hail," repeating the injunction to tell the disciples to go into Galilee, where the Eleven accordingly went into a mountain where Jesus had appointed them, and "when they saw him they worshipped him: but some doubted." This is the whole of Matthew's testimony.

St. Luke, again, in his Gospel and Acts amplifies the miraculous appearances almost up to the extent described by St. Paul, though with considerable differences both of addition and omission. The three women become a number of women; the one angel or young man in shining clothes, two; the appearance to the women disappears; Peter is mentioned as running to the sepulchre but departing without seeing anything special except that the body had been removed; the first appearance recorded is that to the two disciples walking from Emmaus, who knew Him not until their eyes were opened by the breaking of bread, when He vanished; the next appearance is to the Eleven sitting at meat with closed doors; and finally there is the crowning miracle of the Ascension, stated somewhat vaguely in the Gospel, but with more detail in the Acts, describing how He was taken up to heaven and received in a cloud, in the sight of numerous witnesses. This is probably the same miracle as that mentioned by St. Paul as having occurred in the presence of "more than five hundred brethren at once, of whom the greater part remain alive unto this present;" though he mentions two subsequent appearances — one to

James and a second to all the Apostles—of which no trace is found in any other canonical narrative. It is to be noted that all St. Luke's miracles are expressly stated to have occurred at Jerusalem, where Jesus had commanded His disciples to remain, and are, therefore, in direct contradiction with the statements of Matthew and Mark, that whatever occurred was in Galilee, where the disciples were expressly enjoined to go.

When we come to St. John, we find the first part of the narrative of the other Gospels repeated with several variations and a great many additional details. Mary Magdalene is alone and finds the stone removed from the sepulchre. She tells Peter and John, who run together to the tomb; John outruns Peter, but Peter first enters and sees the napkin and linen grave-clothes, but nothing miraculous, and they return to their homes. Mary remains weeping and sees, first two angels, and then Jesus himself, whom she at first does not recognise and mistakes for the gardener. The walk to Emmaus is not mentioned, and the next appearance is to the disciples sitting with closed doors. Another takes place after eight days, for the purpose of convincing Thomas of the reality of the resurrection in the actual body, and here apparently the narrative closes with the appropriate ending, "That these things are written that ye may believe that Jesus is the Christ, the Son of God; and that believing ye might have life through his name." But a supplementary chapter is added, describing a miraculous draught of fishes and appearance to Peter, John, and five other disciples at the Sea of Tiberias in Galilee, in which the command is given to Peter to "Feed my sheep," and an explanation is introduced of what was doubtless a sore perplexity to the early Christian world, the death of St. John before the coming of the Messiah.

These are the depositions of the five witnesses, Matthew, Mark, Luke, John, and Paul, in which the verdict "proven" or "not proven" must rest in regard to the issue "miracle" or "no miracle."

The mere statement of them is enough to show how insufficient they are to establish any ordinary fact, to say nothing of a fact so entirely opposed to all experience as the return to life of one who had really died. Suppose it were a question of proving the signature of a will, what chance would a plaintiff have of obtaining a verdict who produced five witnesses, four of whom could give no certain account of themselves, while the fifth spoke only from hearsay, and the details to which they deposed were hopelessly inconsistent with one another as regards time, place, and other particulars? The account of the Ascension brings this contradiction into the most glaring light. According to St. Luke and St. Paul this miracle took place at Jerusalem, in the presence of a large number, St. Paul says over 500 persons, before whose eyes Jesus was lifted up in the body into the clouds, and more than half, or over 250 of these witnesses, remained alive for at least twenty years afterwards to testify to the fact. Consider what this implies. Such an event occurring publicly in the presence of 500 witnesses is not like an appearance to a few chosen disciples in a room with closed doors: it must have been the talk of all Jerusalem.

The prophet who had shortly before entered the city in triumphal procession amidst the acclamations of the multitude, and who, a few days afterwards, by some sudden revolution of popular feeling, had become the object of mob-hatred; who had been solemnly tried, condemned, and executed; that this prophet had been restored to life and visibly translated in the body to heaven in the presence of more than 500 witnesses, must inevitably have caused an immense sensation. However prone the age might be to believe in miracles, such a miracle as this must have startled every one. The most incredulous must have been converted; the High Priest and Pharisees must, in self-defence, have instituted a rigid inquiry; the Proconsul must have reported to Rome; Josephus, who, not many years afterwards, wrote the annals of the Jews during this period with considerable detail, must have known of the occurrence and mentioned it.

And above all, Matthew, Mark, and John must have been aware of the occurrence; and in all probability, Matthew, John, and Peter, from whom Mark derived his information, must have been among the 500 eye-witnesses. How then is it possible that, if the event really occurred, they not only should not have mentioned it, but partly by their silence, and partly by their statement that they went into Galilee, have virtually contradicted it. The Ascension, if true, was a capital fact, not only crowning and completing the drama of Christ's life which they were narrating with its most triumphant and appropriate ending, but confirming, in the strongest possible manner, the doctrine for which they were contending, that He was not an ordinary man or ordinary prophet, but the Messiah, the Son of God, who had redeemed the world from its original curse and conquered sin and death. One might as well suppose that any one writing the life of Wellington would omit the Battle of Waterloo as that any one writing the life of Christ would knowingly and wilfully omit all mention of the Ascension. It must be evident that whoever wrote the original manuscripts from which the Gospels of Matthew, Mark, and John were compiled, must either never have heard of the Ascension, or having heard of it did not believe it to be true. This must also apply to the other miraculous appearances said to have occurred at Jerusalem. How was it possible for writers who knew of them to make no mention of them, and virtually contradict them by asserting that they did not remain at Jerusalem, but went to Galilee in obedience to a command to that effect, and that the final parting of Jesus from His disciples took place there?

The most unaccountable fact is the total silence of Mark, who was nearest the fountain-head if he derived his information from St. Peter, as to these miraculous appearances. If his Gospel ended with verse 8 of chapter xvi., as the oldest manuscripts and the internal evidence of the postscripts afterwards added appear to prove, there is absolutely no statement of any such appearance at all. Nothing is said but that three women found the tomb empty and saw a young man clothed in white, who told them that Jesus had risen and gone into Galilee. Now, if there is one fact more certain than another about miraculous legends, it is that as long as they have any vitality at all, they increase and multiply and do not dwindle and diminish. We have an excellent example of this in the way in which a whole cycle of miracles grew up in a short time about the central fact of the martyrdom of St. Thomas à Becket.

If, therefore, Matthew and Mark knew nothing of the series of miracles, which from St. Paul's statement we must assume to have been currently believed by the early Christians twenty years after the death of Christ, the only possible explanation is that their Gospels were compiled from narratives which had been written at a still earlier date, before these miracles had been heard of.

We must suppose that Mark really wrote down what he heard from Peter, and that Peter, being a truthful man, though he probably had a sincere general belief that Christ had risen, declined to state facts which he knew had never occurred. This is in entire accordance with what we find in the whole history of ecclesiastical miracles, from those recorded in Scripture down to those of St. Francis of Assisi in the thirteenth century, and of St. Francis Xavier in the sixteenth. Innumerable as are the accounts of miracles said to have been wrought by relics or by other holy persons, there is no instance of any statement by any credible person that he had himself worked a real miracle. St. Augustine describes in detail many wonderful miracles, including resurrections from the dead, which he said had been wrought to his own knowledge, within his own diocese of Hippo, by the relics of the martyr Stephen. In fact, he says that the number of miracles thus wrought within the last two years since when these relics had been at Hippo, was at least seventy. This testimony is far more precise than any for the Gospel miracles, for it comes from a well-known man of high character, who was on the spot at the time, and speaks of these and many other miracles having occurred to his own knowledge. But he never asserts that he himself had ever wrought a miracle.

In like manner Paulinus relates many miracles of his master, St. Ambrose, including one of raising the dead; but Ambrose himself never asserts that he performed a miracle. Neither does St. Francis of Assisi, or any of the 25,000 saints of the Roman calendar to whom miracles are attributed.

Even Jesus himself seems, on several occasions, to have disclaimed the power of working miracles, as when He refused to comply with the perfectly reasonable request of the Jews to attest His Messiahship by a sign, if He wished them to believe in it.

There is every reason, therefore, to believe that when we find narratives making no mention of important miracles which were afterwards commonly received, they must be taken from records of an earlier date, and proceeding directly from those who, if the miracles were true, would have been the principal eye-witnesses to vouch for them. But, if this be so, how near to the fountain-head do these narratives carry us? We lose the miracles, but in compensation we get what may be considered fresh and lively narratives of the life and conversation of Jesus, and confirmation both of His being an historical personage, and of the many anecdotes and sayings which depict His character, and bring Him before us as He really lived. The mythical theory cannot stand which found in every saying and action an *ex post facto* attempt to show that He fulfilled prophecies and realised Messianic expectations. We can see Him walking through the fields on a Sunday afternoon with His disciples, plucking ears of corn, and rebuking the Pharisees for their puritanical adherence to the letter of the observance of the Sabbath; we can see Him taking little children in His arms, and talking familiarly at the well with the woman of Samaria; we can hear Him preaching the Sermon on the Mount, and dropping parables from His mouth like precious pearls of instruction in love, charity, and all Christian virtues. We can sympathise with the agony in the garden as with a real scene, and hear the despairing cry, "My God, my God, why hast thou forsaken me?"

It seems to me that faith in the reality of scenes like these is worth a good deal of faith in the metaphysical conundrums of the Athanasian Creed, or in the actual occurrence of incredible miracles.

Another argument in favour of the early date and genuine character of the primitive records which have been worked up in the Synoptic Gospels, is afforded by the sayings attributed to Jesus. It is impossible to imagine that these could be the invention of a later age, when theological questions of faith and doctrine had absorbed almost the entire attention of the Christian world. We have already seen how wide is the difference, both as regards style and phase of thought, between the discourses reported in the fourth Gospel and those of the Synoptics. No one writing in the second or towards the end of the first century, or even earlier in the religious atmosphere of St. Paul's Epistles, could have composed the Sermon on the Mount or the Lord's Prayer. The parables and maxims, instead of teaching nothing but a pure and sublime morality in simple language, must have contained references to the doctrine of the

Logos, and the disputes between the Jewish and the Gentile Christians. Even if these discourses had passed long through the fluctuating medium of oral tradition, they must, when finally reduced to writing, have shown many traces of the theological questions which agitated the Christian world. The only explanation is that Apostles like St. Matthew, and St. Peter through Mark, really recorded these sayings in writing while they were fresh in memory, and that their authority secured them from adulteration.

At the same time it must be borne in mind that while portions of the original narrative appear to carry us back very near to the fountain-head, a large part of the Gospels in their present form is evidently of much later date and of uncertain origin. It is clear that Papias, writing about the year 150, knew nothing of the Gospels of Luke and John, and nothing of those of Matthew and Mark in their present form. The discourses of Matthew and the disconnected notes of Mark, to which he refers, were something very different from the complete histories of the life and teaching of Jesus contained in the present Gospels. It is equally clear that Justin Martyr and Hegesippus, who wrote about the middle of the second century, and made frequent quotations of the sayings and doings of the Lord, made them, not from the present canonical Gospels, but from other sources relating the same things in different order and different language. "A Gospel according to the Hebrews" and "Memoirs of the Apostles" seem to have been the principal sources from which they quoted.

It is evident however, that during the first two centuries there were a great number of so-called

Gospels and Apostolic writings floating about in the Christian world along with oral traditions. The author of Luke tells us this expressly, and later writers refer to a number of works now unknown or classed as apocryphal, and complain of forged Gospels circulated by heretics. None of these writings, however, seem to have had any peculiar authority or been considered as inspired Scripture, which term is exclusively confined to the Old Testament, until the middle of the second century.

At length, by a sort of law of the survival of the fittest, the present Gospels acquired an increasing authority and superseded the other works which had competed with them; but the selection was determined to a great extent, not by those principles of criticism which would now be applied to historical records, but by doctrinal considerations of the support they gave to prevalent opinions. In other words, orthodoxy and not authenticity was the test applied, and it is probable that no Christian Father of the second or third century would have hesitated to reject an early manuscript traceable very clearly to an Apostle, in favour of a later compilation of doubtful origin, if the former contained passages which seemed to favour heretical views, while the latter omitted those passages, or altered them in a sense favourable to orthodoxy.

To sum up the matter, it appears that while the antecedent improbability of miracles has been enormously increased by the constant and concurrent proofs of the permanence of the laws of Nature, the evidence for them when dispassionately examined, is altogether insufficient to establish even an ordinary fact.

T

CHAPTER IX.

CHRISTIANITY WITHOUT MIRACLES.

Practical and Theoretical Christianity—Example and Teaching of Christ—Christian Dogma—Moral Objections—Inconsistent with Facts—Must be accepted as Parables—Fall and Redemption—Old Creeds must be Transformed or Die—Mahometanism—Decay of Faith—Balance of Advantages—Religious Wars and Persecutions — Intolerance — Sacrifice — Prayer — Absence of Theology in Synoptic Gospels—Opposite Pole to Christianity—Courage and Self-reliance—Belief in God and a Future Life—Based mainly on Christianity—Science gives no Answer—Nor Metaphysics—So-called Intuitions—Development of Idea of God—Best Proof afforded by Christianity—Evolution is Transforming it—Reconciliation of Religion and Science.

CAN Christianity continue to exist without miracles? To answer this question we must distinguish between practical and theoretical Christianity. The essence of practical Christianity consists in such a genuine acceptance of its moral teaching, and love and reverence for the life and character of its Founder, as may influence conduct, and be a guide and support in life. Theoretical Christianity is that which professes to teach a complete theory of the creation of the world and man, of the relations between man and his Creator, and of his position and destiny in a future state of existence.

The former needs no miracles. The Sermon on the Mount, and St. Paul's description of Christian charity, carry their own proof with them, and such parables as that of the Good Samaritan require no support, either from historical evidence or from supernatural signs, to come home to every heart whether in the first or in the nineteenth century. The fact that the son of a Jewish mechanic, born in a small town of an obscure province, without any special aid from position, education, or other outward circumstance, succeeded, by the sheer force of the purity and loveliness of his life and teaching, in captivating all hearts and founding a religion which for nineteen centuries has been the main civilising influence of the world and the faith of its noblest men and noblest races; this fact, I say, is of itself so admirable and wonderful as not to require the aid of vulgar miracles and metaphysical puzzles in order to be recognised as worthy of the highest reverence. And when such a life was crowned by a death which remains the highest type of what is noblest in man, self-sacrifice in the cause of truth and for the good of others, we may well call it divine, and not quarrel with any language or any forms of worship which tend to keep it in view and hold it up to the world as an inducement to a higher life.

Miracles are not only unnecessary for a faith of this description, but are a positive hindrance to it. To put it at the lowest, miracles, in an age which has learned the laws of Nature, must always be open to grave doubts, and thus throw doubt on the reliability of the narratives which are supposed to depend on them. Moreover, the touching beauty and force of example of the life of Jesus are almost lost if He is evaporated into a sort of supernatural being, totally unlike any conceivable member of the human family. We may strive to model our conduct at a humble distance on that of the man Jesus, the carpenter's son, whose father and mother, brothers and sisters, were familiar figures in the streets of Nazareth, but hardly on that of a "Logos," the incarnation of a metaphysical conception of an attribute of the Deity, who existed before all worlds and by whom all things were made.

But, on the other hand, miracles are indispensable for the dogma, or theoretical side of Christian theology. Let us consider frankly what this dogma is, and how far it is *true*—that is, consistent or inconsistent with known and indisputable facts.

The Christian dogma cannot be better stated than in the words of St. Paul, who was its first inventor, or, at any rate, the first by whom it was elaborated into a complete theory.

"For as in Adam all die, even so in Christ shall all be made alive."

This may be expanded into the following propositions:

1. That the Old Testament is miraculously inspired, and contains a literally true account of the creation of the world and of man.

2. That, in accordance with this account, the material universe, earth, sun, moon, and stars, and all living things on the earth and in the seas, were created in six days, after which God rested on the seventh day.

3. That the first man, Adam, was created in the image of God and after His own likeness, and placed, with the first woman, Eve, in the Garden of Eden, where they lived for a time in a state of innocence, and holding familiar converse with God.

4. That by an act of disobedience they fell from this

high state, were banished from the Garden, and sin and death were inflicted as a penalty on them and their descendants.

5. That after long ages, during which mankind remained under this curse, God sent His Son, who assumed human form, and by His sacrifice on the cross appeased God's anger, removed the curse, and destroyed the last enemy, death, giving a glorious resurrection and immortal life to those who believed on Him.

This theory is a complete one, which hangs together in all its parts, and of which no link can be displaced without affecting the others. It is the theory which has been accepted by the Christian world since its first promulgation; and, although expounded with metaphysical refinements in the Athanasian Creed, and set forth with all the gorgeous surroundings of poetical imagination in Milton's "Paradise Lost," it remains in substance St. Paul's theory, that "as in Adam all die, even so in Christ shall all be made alive."

It is obvious that this theory is open to grave objections on moral grounds. It is more in the character of a jealous Oriental despot than of a loving and merciful Father, to inflict such a punishment on hundreds of millions of unoffending creatures for an act of disobedience on the part of a remote ancestor. And it is still more inconsistent with our modern ideas of justice and humanity to require the vicarious sacrifice of an only Son as the condition of forgiving the offence and removing the curse.

Nevertheless it must be admitted that, notwithstanding these objections, and harsh as the theory is, it has had a wonderful attraction for many of the highest intellects and noblest nations of the human race.

It was the creed of Luther, Cromwell, and Milton, and the inspiring spirit of Scotch Presbyterianism and English Puritanism. It has inspired great men and great deeds, and although responsible for a good deal of persecution and fanaticism, it must always be spoken of with respect, as a creed which has had a powerful effect in raising men's minds from lower to higher things, and has on the whole done good work in its time.

But the question of its continuance as a creed which it is possible for sincere men to believe, as literally and historically true, depends not on wishes and feelings, or on reverence for the past, but on hard facts. Is it or is it not consistent with what are now known to be the real truths respecting the constitution of the universe and the origin of life and of man?

To state this question is to answer it. There is hardly one of the facts shown in the preceding chapters to be the undoubted results of modern science which does not shatter to pieces the whole fabric. It is as certain as that two and two make four that the world was not created in the manner described in Genesis; that the sun, moon, and stars are not lights placed in the firmament or solid crystal vault of heaven to give light upon the earth; that animals were not all created in one or two days, and spread over the earth from a common centre in Armenia, after having been shut up in pairs for forty days in an ark, during a universal deluge. And finally, that man is not descended from an Adam created quite recently in God's image, and who fell from a high state by an act of disobedience, but from a long series of Palæolithic ancestors, extending back certainly into the Glacial and probably into the Tertiary period, who have not fallen but progressed, and by a slow and painful process of evolution have gradually developed intelligence, language, arts, and civilisation, from the very rudest and most animal-like beginnings.

Belief in inspiration, the very key-stone of the system, becomes impossible when it is shown that the accounts given of such important matters in the writings professing to be inspired are manifestly untrue; and when the ordinary rules of criticism are brought to bear upon these writings it is at once seen that they are compilations of different ages from various and uncertain sources.

The improbability of miracles is enormously increased by the proof of the uniform operation of natural law throughout the vast domains of space, time, matter, and life; and where the supernatural was formerly considered to be a matter of every-day occurrence, it has vanished step by step, until only the last vestige of it is left in a possible belief in some of the more important and impressive miracles of the Christian dispensation. Even this faint belief is manifestly founded more on reverence for tradition, and love of the religion which the miracles are supposed to support, than on any dispassionate view of the evidence on which they rest. Tried by the ordinary rules of evidence, it is apparent that it is contradictory and uncertain, and not such as would be sufficient to establish in a court of law any ordinary fact, such as the execution of a deed. It is apparent also that the evidence for the most crucial and important of all miracles, that of the Ascension, is not nearly so precise and cogent as that for a number of early Christian and mediæval miracles which we reject without hesitation.

What follows? Must we reject these venerable traditions as old wives' fables? I answer, No; but we must accept them as parables.

A great deal of the best teaching of the New Testament is conveyed in the form of parables. Take for instance that of Lazarus and Dives. No one supposes that this is an historical narrative; that this particular Jew, out of the millions of poor and good Jews who have lived and died, was actually taken up into Abraham's bosom; and that the remarkable dialogue across the gulf is a literal transcript of an actual conversation. But the moral is taught for all time, that it is bad for the rich to indulge in selfish luxury and take no thought of the mass of poverty and misery weltering around them; and that the condition of the poorest of the poor, borne with piety and resignation, may really be better and higher than that of the selfish rich. Apply the same principle to the dogma of the fall and redemption, and we may see in it a parable of the

NOTE.—Since writing this chapter, I have seen with much pleasure an article entitled "Christmas," by Matthew Arnold, in a recent number of the *Contemporary Review*, which takes exactly the same view of the allegorical or parabolic sense of miraculous narratives. He takes the instance of the Immaculate Conception and Birth of the Saviour, and shows that it was a myth which grew up, almost inevitably, from the strong impression made on the minds of early Christians by the idea of purity set forth by the life and teaching of Jesus, which stood in such striking contrast with the corruption of the heathen world. The same idea led to a similar myth in the case of Gautama, the pure and self-sacrificing founder of the Buddhist religion, and it teaches an eternal truth to all who can look below the letter to the spirit of the parable.

highest meaning. Every one of us must be conscious of having fallen by yielding to temptation and giving way to animal passions. We may have fallen so low that without some redemption, or friendly influence from without, we cannot raise ourselves from the lower level and regain our lost place. We can see that there are thousands round us, who, from poverty or other adverse circumstances, have got immersed in evil conditions from which it is hopeless to extricate themselves without friendly aid. We can see also that there is nothing more noble and divine than to make sacrifices in order to be the redeemer who saves as many souls as possible from this entanglement of evil, and gives them a chance of rising into a happier and better life. We may feel this, and use as an incentive to attempt some humble imitation of it, the parable which presents it to us in its highest aspect, and has been the efficient means of stimulating so many good men to do good works. This is surely better than paltering with the truth, and enervating our conscience and intelligence by professing to believe in the literal historical accuracy of things which have become incredible to all thinking and educated minds. Of course, I do not mean that these dogmas and miraculous narratives were intended by the original writers to be parables, but only that they have become so to us; and the alternative lies between rejecting them altogether or accepting them as having an allegorical meaning or latent truth.

At any rate, whether we like it or not, this is what we shall have to do, for the conclusions of science are irresistible, and old forms of faith, however venerable and however endeared by a thousand associations, have no more chance in a collision with science than George Stephenson's cow had if it stood on the rails and tried to stop the progress of a locomotive. It is not enough to say that a thing is lovely and amiable, and that its loss will leave a blank, to ensure its continuance. The law of Nature is progress and not happiness. Stars, suns, planets, human individuals, and human races have their periods of youth, maturity, and decay, and are continually being transformed into new phases.

> The old order changes, giving place to new,
> And God fulfils Himself in many ways.

Childhood, with its innocence and engaging ways, passes into the sterner and more prosaic attributes of the grown-up man; fancy decays as reason ripens; simple faith is replaced by larger knowledge; and the smooth brow of infancy is often marred by wrinkles of strife and suffering, impressed during the more or less successful struggle in the battle of life; and yet we could not if we would, and would not if we could, arrest the progress of Nature, and say that the child shall never grow into a man.

Such also is the fate of creeds. They must be transformed or die; and the best test of the vitality and intrinsic truth of a religion is just that capacity for transformation against which theologians exclaim as sacrilege. In this respect Christianity has a great advantage over other religions. The pious souls who are shocked at any denial of the inspiration of Scripture may console themselves by considering what has been the fate of other religions which have been imprisoned too closely within the limits of a sacred book. Mahometanism, the religion of one God and a succession of prophets or great men who have taught his doctrines, is not in theory inconsistent with progress and civilisation. But Mahomet unfortunately wrote a book, the Koran, which, while it contained much that to the Arab mind was sublime and beautiful, was of necessity impregnated with the ideas of the age in which he lived; an age of much ignorance and superstition, of imperfect social arrangements, and of barbarous and ferocious manners. This book came to be accepted as the inspired word of Allah, which it was impious to question, to which nothing could be added, and from which nothing could be taken away. Hence Mahometanism has become what we see it — a narrow and fanatical creed, incompatible with progress and free thought, and stereotyping institutions, such as polygamy and slavery, which are fatal to any advance towards a higher civilisation. From this fate Christianity has been saved by the fortunate circumstance that its sacred books are collections of a variety of writings of different authors and different ages, reflecting such various and often conflicting phases of thought and belief that of necessity their interpretation was very elastic, and lent itself readily to the changes required by the spirit of successive periods and of different nationalities. Wherever for a time a system of infallibility was enforced, as in Spain by the Inquisition, Christianity became cruel, barbarous, unprogressive, and really very little better than the religion of Islam, to which it closely approximated. Decay of faith, therefore, in dogmatic Christianity is, like other great revolutions of thought, a question, not of absolute gain or absolute loss, but of a balance between conflicting advantages and disadvantages.

The loss is evident enough, and is set forth with much eloquence and force by the few remaining champions of orthodoxy. The simple, undoubting faith, which has been for ages the support and consolation of a large portion of mankind, especially of the weak, the humble, and the unlearned, who form an immense majority, cannot disappear without a painful wrench, and leaving, for a time, a great blank behind. But, on the other hand, there are a great many real and important advantages which have to be set on the credit side of the account.

Intolerance is the shadow which dogs the footsteps of faith, and in many cases more than obscures its benefits. When we consider the mass of human misery which has been occasioned by religious wars and persecutions; the ruthless extirpation of the Albigenses; the slaughter of the saints

whose bones
Lie scattered on the Alpine mountains cold;

the Thirty Years' War, which desolated Germany and threw civilisation back for a century; the civil wars of France; the Spanish Inquisition; and a thousand other instances of the baleful effects of religious hatreds, we can almost sympathise with those who pronounce religion an invention of priests for the promotion of evil, and exclaim with the Roman poet:

Religio tantum potuit suadere malorum.

To this must be added the misery caused by the belief in demonology and witchcraft, and the tortures inflicted on innumerable innocent victims by prejudices inspired by a literal construction of passages of the Old Testament. Nor is it a small matter to have escaped from the nightmare dreams which must have oppressed so many minds, especially of the young and imaginative, in an age when such a book as Dante's "Inferno" could be written, and accepted as a gleam of prophetic insight into the horrors of the invisible world.

Even in more recent and humane times, intolerance remained as a general mode of thought, inspiring hatred of those whose form of belief differed from that which was generally adopted. It is only within the present generation that true tolerance has come to be established as the law of modern thought, and men have learned to live together and love one another, without reference to intellectual differences of creed and doctrine. Surely this is a great advantage, and we are nearer to the true spirit of Christianity than in the days when a Birmingham mob sacked Priestley's house because he professed his belief in the saying of Jesus, that "my Father is greater than I." We may read the Athanasian Creed less, but we practise Christian charity more, in the present than in any former age.

Another great advantage is that as freer thought has been brought to bear on the mysteries of religion, we have purged off the grosser ideas and arrived at much more enlarged and spiritual conceptions. Take, for instance, prayer and sacrifice. In its crude form, sacrifice was a sort of bargain struck with an unseen Power, by which we hoped to obtain some favour which we greatly desired, in exchange for giving up something which we greatly valued. This is the form in which sacrifice appears in the Old Testament, in Abraham's offer to kill his son Isaac, and in the record of the Moabitish stone, how the king, when besieged in his capital, sacrificed his son, and by so doing obtained the favour of his God and defeated his enemies. In another form, sacrifice was considered as a propitiation to appease the anger of an offended Deity, pictured as a sort of Oriental despot, who must have some one for a victim, and was not particular who it might be; and even in the Christian dogma the merit of the sacrifice is very closely analogous to that of the Mayor of Calais who went out to King Edward with a halter round his neck, ready to be hanged, so that he might save the lives of his fellow-citizens.

Nowadays, no one thinks of sacrifice as anything but the sacrifice of lower instincts and passing temptations to a higher ideal, and the voluntary renunciation of selfish ease and pleasure for the good of others.

In like manner, the original idea of prayer was that of obtaining a request by flattery or importunity, just as a courtier might do at the court of some earthly king of kings or sultan. It is now spiritualised into the conception that its effect is entirely subjective; that it never really obtains any reversal of the laws of Nature, but that it often exalts the mind to a frame in which things otherwise impossible become possible. A German regiment marches to battle singing Luther's grand old hymn—

Ein feste Burg ist unser Gott.

Half the regiment may be freethinkers, but it is nevertheless true that they are more likely to stand firm and win the victory if they chant the hymn, than if they march in silence.

Taking all these things into account, there is no reason to despair because the irresistible progress of science has made us

Falter where we firmly trod,

and changed a great deal of what was once fixed and certain faith into vague aspirations and less definite, though larger and more spiritual, conceptions.

There is next to no theology in the Christianity of the Synoptic Gospels, which give us by far the nearest and most authentic record of what its Founder actually taught; and it may be that in sloughing off the mythical legends and metaphysical dogmas which have grown up around it, we shall be, in reality, not banishing the Christian religion from the world, but making it revert to its more simple and spiritual ancestral type, in which form all that is really valuable in its pure and elevated morality may be incorporated more readily with practical life, and assimilated without difficulty with the progressive evolution of modern thought and science.

At the same time we must bear in mind that even Christianity in its purest form does not escape from the universal law of polarity, and presents, not the whole truth, but only one very important side of truth. It is the religion of love, purity, gentleness, and charity; important virtues, but not all that constitute the perfection of men or nations. In fact, if carried to the "falsehood of extremes," its very virtues become vices. It would not work in practice, if smitten on one cheek to turn the other; and any one who attempted to follow literally the precept of "taking no thought for the morrow," and trusting to be fed like the sparrows, would, in modern society, come dangerously near being what we call in Scotland a "ne'er-do-weel," that is to say, a soft, molluscous sort of creature, who is a burden on his friends, and ends his days as a pensioner on charity or a writer of begging letters. The foremost men and foremost races of modern society are precisely those who act on the opposite principle, and do look ahead and steer wisely and boldly amidst dangers and difficulties for distant and definite ends.

In one of the old Norse Saga there is a saying which has always impressed me greatly. An aged warrior, when asked what he thought of the new religion, replied: "I have heard a great deal of talk of the old Odin and of the new Christ, but whenever things have come to a real pinch, I have always found that my surest trust was in my own right arm and good sword."

This strong self-reliance and hardy courage to do or to endure is, beyond all doubt, the solid rock foundation upon which the manly character of individuals and of nations must be built up. The softer virtues and graces come afterwards, which are to refine and adorn, and convert the man into the *gentle* man, or one of Nature's true gentlemen. But without the harder gifts of courage and self-help, a man is not a man, and the raw material is not there out of which to fashion a Gordon or Christian hero.

This may be called the Norse pole as contrasted with the pole of Christianity, and the perfect man is he who can stand firmly between the two opposites, controlling both and being controlled by neither.

While I have thought it right, however, to call attention to this counter-pole to Christianity, I should add that with the strong, practical Teutonic races there is not much danger of erring on the side of too much weakness, humility, or asceticism, and therefore the influence of the Christian religion makes mainly for good. Modern civilisation has been formed, to a great extent, by grafting the gentler virtues of the Gospel on the robust primitive stock of the barbarians who overthrew Rome. It is the example and teaching of Jesus, the son of the carpenter of Nazareth, which have been mainly instrumental in diffusing ideas of divine love, charity, and purity throughout the world, and humanising the iron-clad and iron-souled warriors, whose trust was in their stout hearts and strong right arms, and who knew no law but

> The good old plan,
> That he should take who has the power,
> And he should keep who can.

In another respect it is most important that the world should, as far and as long as possible, hold on to Christianity and struggle to save its essential spirit from the shipwreck of its theology, and the sheer impossibility of believing in the literal and historical truth of many of its dogmas.

The highest and most consoling beliefs of the human mind are to a great extent bound up with the Christian religion. If we ask ourselves frankly how much, apart from this religion, would remain of faith in a God and in a future state of existence, the answer must be, very little. Science traces everything back to primeval atoms and germs, and there it leaves us. How came these atoms and energies there, from which this wonderful universe of worlds has been evolved by inevitable laws? What are they in their essence, and what do they mean? The only answer is, it is unknowable. It is "behind the veil," and may be anything. Spirit may be matter, matter may be spirit.

We have no faculties by which we can even form a conception, from any discoveries of the telescope or microscope, from any experiments in the laboratory, or from any facts susceptible of real human knowledge, of what may be the first cause underlying all these phenomena.

In like manner we can already to a great extent, and probably in a short time shall be able to the fullest extent, to trace the whole development of life from the lowest to the highest; from protoplasm, through monera, infusoria, mollusca, vertebrata, fish, reptile, and mammal, up to man—and the individual man from the microscopic egg, through the various stages of its evolution up to birth, childhood, maturity, decline, and death. We can trace also the development of the human race through enormous periods of time, from the rudest beginnings up to its present level of civilisation, and show how arts, languages, morals, and religions have been evolved gradually by natural laws from primitive elements, many of which are common in their ultimate form to man and the animal creation.

But here also science stops. Science can give no account of how these germs and nucleated cells, endowed with these marvellous capacities for evolution, came into existence or got their intrinsic powers. Nor can science enable us to form the remotest conception of what will become of life, consciousness, and conscience, when the material conditions with which they are always associated while within human experience, have been dissolved by death and no longer exist. We know as little in the way of accurate and demonstrable knowledge of our condition after death as we do of our existence—if we had an existence—before birth.

If we turn for an answer to these questions from science to metaphysics we find ourselves in cloud-land. Mists of fine phrases and plausible conjectures form into philosophies, and dissolve away again without leaving a vestige of positive knowledge. Take Descartes' famous fundamental axiom, "Cogito, ergo sum," I think and therefore I exist. Is it really an axiom? Does it take us any nearer to what thought really is, and what is the true meaning of existence? If the fact that I am conscious of thinking proves the fact that I exist, is the converse true, that whatever does not think does not exist? Am I existent or non-existent during the seven or eight hours of dreamless sleep out of every twenty-four, when to a certainty I am not thinking? Does a child only begin to exist when it begins to think? If "Cogito, ergo sum" is an intuition to which we can trust, why is not "Non cogito, ergo non sum" an equally good foundation on which to build a system of philosophy, and spin out of the brain an ideal system of God, man, and the universe?

The so-called intuitions of metaphysics seem really to amount to little more than translations into philosophical language of our own earnest wishes and aspirations. We shudder at the notion of annihilation; we revolt at the idea that all the high faculties of the mature and cultivated mind are to be extinguished by death; we long for a future life, in which we may again see beloved faces, and, pondering on these things, we have a strong impression that it must not and cannot be, which presently takes the form, in some minds of a philosophical turn, of what is called an intuition, on which they proceed to build up a demonstration of God and immortality.

But, again, what do they really know more than science has already told us? The essence of all spiritual existence, as far as we know anything of it, is personal consciousness. This clearly depends on, or is indissolubly associated with, a certain condition of a material organ—the brain. With a less active condition of this organ, as in sleep, personal consciousness is suspended. In the case of a man recovered from drowning by artificial means, it is gone, and the man is to all intents and purposes dead for perhaps a quarter of an hour, and would remain dead if warm blankets and artificial respiration did not recall him to life. Where and what was he during this interval? and, if his personal identity and conscious existence were gone for that quarter of an hour, why and when did they return? and, if the Humane Society's men had been less prompt, would they ever have returned?

These are questions to which no metaphysical system that I have ever seen can return the semblance of an answer.

Again, how is it possible for philosophy to lay down as an axiom that man has an intuitive perception of a Deity, in the face of the fact that whole races of savage men have no such perception, and have not got beyond rude fetichism and a vague superstitious fear of ghosts and evil spirits, while others, further advanced, have made their own anthropomorphic gods, obviously from reflections of their own faculties and passions on the distant mists of the unknown, like the spectres of the Brocken? We can trace the idea of Deity, step by step, from early attempts to explain phenomena of nature, astronomical, legendary, and linguistic myths, and reverence for departed ancestors and heroes, up to the philosophical conceptions of a Plato or a Marcus Aurelius. In the same way we can trace, step by step, the transformation of the tribal God of Abraham, Isaac, and Jacob, into the national God of Israel, who was at first only better and stronger than the gods of the surrounding nations, but finally became the sole God of the universe, degrading the other gods to the category of dumb idols. So, also, we can see the first crude anthropomorphic conceptions of this Deity gradually giving way to purer and nobler ideas. The God who required rest on the seventh day becomes the Almighty one at whose word all things were created. The jealous and cruel God who withdrew His favour from the chivalrous Saul, because he would not hew his captives in pieces before the Lord, is transformed into the God who "loves mercy and not sacrifice." The God who found after His own heart the man whose depraved mind could conceive such an act of foul villainy as David practised towards Uriah, and who not only condoned the crime, but rewarded it by giving the succession to the son of the adulterous intercourse with Bathsheba, has become the God of holy love and purity of the New Testament. At which of these stages did the philosophical intuition of God come in, which is said to be an innate faculty of the human mind, and the surest base of all our knowledge of the universe? Where is the inevitable intuitive perception of a personal Deity in the minds of some of the deepest thinkers and purest livers of the present day, who, like Herbert Spencer, can discern nothing behind the veil but a great unspeakable and unknowable?

After all we must fall back on Christianity for any grounds upon which to trust, more or less faintly, in the "larger hope." The Christian religion, apart from any question of miracles, is an existing fact. It is a fact which for nineteen centuries has proved, on the whole, in accordance with other facts and with the deepest feelings and highest aspirations of the noblest men and women of the foremost races in the progressive march of civilisation. Why do we say that its moral teachings, such as we find in the Sermon on the Mount, and in St. Paul's definition of Christian charity, carry conviction with them and prove themselves? Because they accord with, and give the best expression to, feelings, which in the course of evolution of the human mind from barbarism to civilisation have become instinctive. We may be able to trace their origin and development, we may be able to see that they are not primary instincts, implanted at birth like those of the lower animals, but secondary instincts, formed by the action of a civilised environment on hereditary aptitudes. Still there they are, and being what they are, and living in the age and society in which we actually live, they are inevitable and necessary instincts, and it requires no train of reasoning or laboured reflection to make us feel that "right is right," and that it is better for ourselves and others to act on such precepts as those of "loving our neighbours as ourselves," and "doing as we would be done by," rather than to reverse these rules and obey the selfish promptings of animal nature. Of the same order, though less clear and cogent, are the teachings of the Gospel respecting God and immortality. They are less clear and less cogent, because the only evidence by which they could be demonstrated from without, that of miracles, has broken down and failed us; and because we cannot verify them experimentally by an appeal to facts, as we can in regard to the working of moral laws and precepts. But it still remains that they are ideas which have arisen inevitably in the course of the evolution of the human mind; and that they fit in with and satisfy, in a way which no other ideas can do, many of the best and deepest feelings which have equally been developed in that mind, in the course of its progressive ascent from lower to higher things. It remains also that true science, while it can add nothing to this proof, takes nothing from it, and while it excludes miracles and supernatural interference after the order of the universe has been once established, leads us back step by step to a great Unknown, in which, from the very fact that it is unknown, everything is possible.

Further than this it is not possible to carry the proof. If we are to believe at all in a God, we must be content to believe that He knows better than we do what is right and consistent with the conditions of our own existence and that of the universe; and that part of the scheme is, that at a certain stage of the development of our race we should have to exchange the certainty of simple and limited faith for the fainter trust in a larger hope. We may, perhaps, dimly discern something analogous in the progress of each individual from childhood to manhood. He has to part with many a simple belief and unhesitating trust, and climb the hill of life staggering under many a burden of doubt and difficulty; and yet it is better for him to "set a stiff heart to a steep brae," and struggle upwards while life is in him, rather than to remain an innocent child playing at its foot.

Anyhow, whether we like it or not, this is the fact we have to accept; but the hill is steep, the burden heavy, and we may well be grateful to anything which, however vaguely, helps and cheers us on the way; and from this point of view, the ideas of God and of a future life taught by the Christian religion, accepted by so many good men, and hallowed by so many venerable traditions and sacred associations, should be cherished, as far as it is possible to do so without shutting our eyes to facts and indulging in conscious insincerity.

For the same reason we shall do well to be tender with the forms and creeds of this religion, even when they appear to be getting obsolete, and their strict and literal interpretation no longer consistent with known truths. It is far better that the transformation requisite to bring them into accordance with the evolution of modern thought caused by the discoveries of science, should take place gradually and spontaneously from within, rather than forcibly and abruptly from without. Evolutionists specially ought to trust to the healing influences of time, and the inevitable though gradual survival of that which is most in harmony with its existing environment.

Already a great deal has been silently done in this direction. Intolerance and fanaticism have almost disappeared from all cultured minds. Even in the ranks of the clergy themselves, many, in all denominations, are devoting themselves more and more to good works, and less to theological disputes and sectarian wranglings.

The metaphysical side of Christian dogma is fast receding into the far distance. The Athanasian Creed, which once convulsed empires and occupied a foremost place in the thought of the age, has become a mere form, read once or twice a year by lukewarm preachers to indifferent or scandalised audiences, who would be only too glad to have a decent excuse for dropping it out of sight altogether. Let any sincere Christian put to himself candidly the question what part the "Holy Ghost," or the definition of the "Logos," really has in the living faith which guides his actions, and he will be astonished to find into what infinitesimal proportions these once vital dogmas have actually faded. It will be the same with all dogmas which, in their literal and historical interpretation, contradict established facts. They will be either forgotten, or, if they contain a kernel of spiritual meaning, will be transformed into truths taught by parables.

In the meantime, it behoves those who see more clearly than others the absolute certainty of the conclusions of science, and the inevitably fatal results to religion of staking its existence on literal interpretations which have become flatly incredible, to do their best to assist the transformation of the old dogmatic theology into a new "Christianity without miracles," which shall retain the essential spirit, the pure morality, the consoling beliefs, and as far as possible the venerable forms and sacred associations of the old faith, while placing them in thorough accordance with freedom of thought, and with the whole body of other truths, discovered and to be discovered, respecting the universe and man.

CHAPTER X.

PRACTICAL LIFE.

Conscience—Right is Right—Self-reverence—Courage—Respectability—Influence of Press—Respect for Women—Self-respect of Nations—Democracy and Imperialism—Self-knowledge—Conceit—Luck—Speculation—Money-making—Practical Aims of Life—Self-control—Conflict of Reason and Instinct—Temper—Manners—Good Habits in Youth—Success in Practical Life—Education—Stoicism—Conclusion.

> Self-reverence, self-knowledge, self-control,
> These three alone lead life to sovereign power.
> Yet not for power; that of itself
> Would come uncalled for; but to live by rule,
> Acting the rule we live by without fear,
> And because right is right to follow right,
> Were wisdom in the scorn of consequence.
>   TENNYSON, *Œnone*.

IN these lines, which he puts into the mouth of the goddess of wisdom, Tennyson, the same poet who has already condensed the essence of modern thought in the lines already quoted from "In Memoriam," gives us what may be well called "the Gospel of practical life." It is clearly our highest wisdom to follow right, not from selfish calculation or hope of reward, but because "right is right;" in other words, because we have a standard within us which tells us, in an unmistakable voice, what to do and what to refrain from doing. For practical purposes, it is comparatively unimportant how this standard got there; whether, according to old creeds, by direct inspiration, or, as modern science tells us, by the slow evolution of primitive faculties, and the accumulation through countless generations of hereditary influences tending towards the survival of the fittest, both of individuals and of societies, in the struggle for life. In either case the standard is there, not as a vague and theoretical, but as an absolute and imperative rule, and the difficulty is not to discern it, but to act up to it.

It may be that it is to a great extent the product of education, and depends on the environment in which we are brought up. It may be that if I had been kidnapped when a child by Comanche Indians, I should have grown up with a very different moral standard touching the taking of scalps and the practice of treacherous murder. But I have not been so kidnapped, and having been born and brought up in a civilised country of the nineteenth century, it is inevitable that outward influences combined with inward capacities should give me a conscience, which tells me in clear enough accents whether I am doing right or wrong. And it is equally certain that by acting in accordance with this conscience, I shall, on the whole, be doing better for myself and better for others than by disregarding it. It is none too easy to make our life even a tolerable approximation towards doing right for the sake of right, and it would be folly to allow any theoretical considerations as to the origin of the idea of right to be an excuse for relaxing any of the constant and strenuous effort which is requisite to keep our feet from straying from the straight path. It is much wiser to cast around us for influences and inducements to strengthen the inward law, and to endeavour by clear insight to bring reason to the aid of faith, and enable us to see intelligently the main causes both of our weakness and of our strength.

This is what the poet does for us in the lines above quoted. Rightly considered, "self-reverence, self-knowledge, and self-control" are the three pillars which support the edifice of a wise and well-ordered practical life.

Self-reverence, in its widest meaning, includes the faculty of forming some ideal standard superior to the lower nature of animal man, and recognising in ourselves some power of approximating to it. The higher the standard the nobler will be the man who cherishes it and tries to attain to it, but it is by no means a rare gift confined to a few select natures. On the contrary, it is the commonest and most universal incentive to good conduct. Even in the rudest and simplest form of admiration for physical courage, it makes heroes of many a common soldier and sailor. If poor Tommy Atkins, fresh from the plough-tail, stands firm in the shattered squares of Waterloo, or on the bloody ridge of Inkermann, it is because he has been brought up in the fixed idea that a Briton must not run away from a Frenchman or a Russian.

In civil life the idea of respectability, though not a very elevated one and apt to degenerate into narrowness, and what Carlyle and Arnold sneer at as "Gigmanity" and "Philistinism," is yet one of very universal and, on the whole, beneficial influence. A large majority of the middle and upper working classes lead decorous lives very much because they feel it incumbent on them to be "respectable," in their own eyes and those of their neighbours. In the case of one half of the human race, the female half, the feeling of self-respect and the desire to be what is called respectable afford the strongest and most constantly present securities both for good morals and good manners. The immense majority of British women are modest maidens and faithful wives, not so much from any cold calculation of the balance of advantages, or from fear of consequences, as from an instinctive feeling that they cannot be otherwise without losing caste and forfeiting their own self-respect and that of their neighbours.

From these common and universal forms of "self-reverence" we rise, step by step, to the higher ideals, which, in every rank and every condition of life, give us among gifted natures what may be called the "salt of the earth," and the shining examples which guide the world to higher things—noble men and noble women. A Sidney, dying on the field of Zutphen, hands over the cup of water to a wounded soldier because his soul, nourished on noble thoughts, and his fancy, fed by the old ballads which, like that of "Chevy Chase," stirred him like a trumpet-blast, had led him to conceive an ideal of a perfect knight which would have been tarnished by any shade of a selfish action. Gordon sacrifices his life at Khartoum, not only cheerfully but almost instinctively, because the suggestion that he might save himself by abandoning those who had trusted in him seems an absolute impossibility.

It is a great advantage of the present day that

education and the press bring such instances of devoted heroism vividly before millions who would never otherwise have heard of them. The influence of the press, both in the way of books and newspapers, is happily in this country almost entirely one which makes for good. There is not a noble act done throughout the world, by high or low, by private or officer, by soldier or civilian, which is not held up for praise and admiration; while any signal instance of cowardice or selfishness is held up to contempt. Newspaper correspondence and leading articles have, to a great extent, superseded sermons, and do the practical moral work of the world in asserting the right and rebuking wickedness in high places. In like manner all the higher works of poetry, fiction, and biography have a good tendency, and are read by an ever-increasing number of readers. Enid and Elaine, Jeanie Deans, Laura Pendennis, Lucy Roberts, are the sort of models set before girls; while boys who have any heroic fibre in their nature are fed with such lives as those of Lawrence and Gordon. For all, but especially for the young, there is no help to self-improvement so great as to read good books in a generous spirit; and nothing which dwarfs the mind so much as to debauch it by frivolous reading, and by the moral dram-drinking of sensational rubbish, until it loses all natural and healthy appetite for the pure and elevated. An affectation of narrow knowingness is also a very fatal tendency in the youthful mind. A man from whose mouth such words as "rot" and "humbug" are constantly heard is, in nine cases out of ten, a very poor, rotten creature himself.

Among the many advantages of self-respect, not

the least important is that it teaches respect for others. The petty jealousies and suspicions, the senseless quarrels, the slanderings and backbitings, which so often turn sour the wine of life, disappear of themselves when a proper standard of self-respect has been firmly established, and a high ideal of human life has become part of our nature. As Tennyson says:

> Like simple noble natures credulous
> Of what they wish for, good in friend or foe;

while on the other hand

> The long-necked geese of the world
> Are always hissing dispraise, because their natures are little.

There are some who delight in running down everything and everybody, and whose appetite for scandal is so great that they are positively unable to refrain from believing and spreading an ill-natured tale, if it affects some eminent man, and still more if it affects a well-known woman. Such are assuredly not the sort of persons whom we should like to resemble ourselves, or to see our sons and daughters resemble. I have always found through life, a safe rule to go by was, if you hear an ill-natured story of a man, discount nine-tenths of it as a lie, and if of a woman, don't believe a word of it.

Perhaps the best test of the amount of real "self-reverence" in an individual or a nation, is to be found in the tone and manner in which women are treated. A low tone invariably bespeaks a low nature, and testifies to innate coarseness and snobbishness, however high may be the rank and polished the outward varnish of the person who indulges in it. On the other hand the roughest miner or backwoodsman is already more than half a gentleman, if his attitude towards women is one of chivalrous courtesy. Nothing looks more hopeful for the future of the human race than to see that the female half of it are constant gainers by the progress of freedom and education. It goes a long way to reconcile one to the dangers of democracy, to find that in the newest and most democratic countries of the world, such as the United States and British colonies, women can travel alone without fear of insult, and have far more innocent liberty and freedom of thought and action than they have in older societies. Whatever may be the case as regards men, for women there can be no doubt that there is a progressive scale upwards from East to West, from despotism to freedom, from Turkey to America.

What has been said of individuals is even more true of nations. Self-respect is the very essence of national life. A great nation may suffer great disasters, and survive them, if the spirit of its people remains intact. England survived the war of American independence, and Prussia recovered from the defeat of Jena. But if a nation loses its vigour and self-respect, if it begins to groan under the burdens of extended empire, and to prefer comfort to honour, ignoble case to noble effort, the hour of its decline has sounded. Imperial Rome did not long survive when she began to contract her frontiers and buy off barbarians. The most fatal thing any Government can do for a country is to destroy its sense of self-respect and teach it to acquiesce in what is felt to be dishonourable.

Looking forward to the future of the great British Empire, this is evidently a turning-point of its destinies.

The triumph of democracy is an inevitable fact; for knowledge is power, and whether for good or evil, the masses have either acquired, or are fast acquiring knowledge, and with equal political rights numbers will tell. How will this democracy of the future affect Imperial interests, and what will be its attitude in regard to foreign and colonial policy?

On the one hand it may be hoped that by making our institutions more popular, and going down to the heart of the masses, our policy will acquire fresh energy and our public men fresh vigour. The working classes are very patriotic, and, on the whole, more open to the influence of generous ideas than the class immediately above them. In the recent instance of the great civil war in the United States, we have seen a democracy making greater sacrifices of men and money for the idea of maintaining national greatness, than was probably ever voluntarily made by any monarchical or aristocratic country. The Copper-heads, who preached peace where there was no peace, and advised letting the erring sisters go their way rather than spend lives and money in the attempt to coerce them, found no response from a nation who felt that the union was their union, and its greatness the separate personal possession of each individual citizen.

But, on the other hand, demagogues will never be wanting to flatter the people, and angle for power by appealing to their lower instincts and advocating measures of present ease and popularity. If a necessity arises for maintaining by the sword an empire which has been won by the sword, the army of parochial politicians who gauge everything by the standard of pounds, shillings, and pence, will be reinforced by the far more respectable body of sentimentalists and humanitarians, who shrink from the shedding of blood in wars the abstract justice of which is not absolutely demonstrated. A large number, perhaps a majority, of platform orators will therefore be found now, as it was in the days of Demosthenes, to denounce armaments, ridicule precautions, minimise responsibilities, and look upon India, the Colonies, and extended empire generally, as troublesome encumbrances rather than as glorious possessions. The two conflicting ideals constantly set before our future political rulers, the four millions whose votes decide the fate of policies and of ministries, will be, on the one hand, that our first duty is to hand down the British Empire to our sons no less great and glorious than we received it from our fathers; on the other, that it is better to stay at home, mind our own affairs, avoid entanglements, contract responsibilities, pass reform bills, and reduce taxes, trusting to the "silver streak" and the chapter of accidents to protect us from invasion. It is the old story of the fable of Hercules, which presents itself constantly to each individual and to every nation. Shall we follow the strait and narrow path which leads upwards, or the broad and easy one which leads, with a pleasant slope to a lower level? Would it have been better for Paris to give the golden apple to Minerva, counselling "self-reverence, self-knowledge, self-control," or to Venus, promising pleasure?

SELF-KNOWLEDGE.

> Oh wad some Power the giftie gie us
> To see ourselves as ithers see us!
>
> BURNS.

A gift which is unfortunately as rare as it is necessary. Without self-knowledge to see our faults how shall we correct them? How shall we become wise if insensible to our follies? How shall we achieve success if we learn no lessons from our failures? There are some men so blinded by vanity that they go through life committing ungentlemanly actions while fancying themselves perfect gentlemen; who are convinced that all men admire them and all women are in love with them, while in reality every one sees through them and laughs at them. A thoroughly impervious vanity is like a waterproof, which throws off the wholesome rain on the outside, while on the inside it is soaked with unhealthy exhalations.

Fortunately this type of vanity is not a common one with our English race, who are too proud and self-reliant to feel the petty anxiety of the really vain man to be always shining in the eyes of others. With us it takes more the form of priding ourselves on artificial distinctions, and attaching an exaggerated importance to matters of trivial importance. Your commonplace English swell, for instance, is apt to class all mankind under two categories—those who associate with lords and wear clothes of a fashionable cut, and those who do not, and to set down all the former as the "right sort," and all the latter as "brutes."

It is a sign of narrowness to make a fetich of these or any other arbitrary distinctions between an upper ten and the rest of mankind, and self-knowledge is never more required than to show the hollowness of adventitious advantages which are not supported by intrinsic merit. A true gentleman feels

> The rank is but the guinea stamp,
> The man's the gowd for a' that,

and feeling this, he holds out the hand of hearty human sympathy to peasant as well as to peer. If born to rank and riches, self-knowledge tells him that he is simply placed on a pedestal, where, if he fails to act on the maxim that "noblesse oblige," the failure will be the more conspicuous. No man who really knows himself can ever be conceited, for he must be aware how far he has fallen short in practice of his own ideal standard, and how constantly "he has done things he ought not to have done, and left undone things he ought to have done."

On the other hand, there is an opposite extreme from which self-knowledge will save a man: that of undue despondency and want of proper confidence and self-reliance. There are men who fail in everything they undertake because they have not the heart to undertake it resolutely, and who at last sink down into the hopeless condition of querulous mental invalids, who cherish their ailments rather than combat them, and are rather proud than otherwise to be considered as interesting victims of untoward circumstances.

For all the relations of practical life the one essential requisite of success is to see things as they really are, and not as we wish them to be; and for this purpose self-knowledge is the foundation of clear insight. If the focus of the glass is wrongly adjusted it will show only distorted images, but if a clear eye looks through a properly focussed glass, outward objects will be truly represented.

Perhaps the commonest of all delusions is that of being born under a lucky star. A man gambles, bets, or speculates because he thinks he is lucky and sure to win. Now, there is in reality no such thing as luck, it is all a question of averages. The only approach to what may be called luck is, that a fool will probably have more of it than a wise man, for as the fool foresees nothing, whenever fortune's die turns up in his favour he sets it down to luck, while the wise man, who has schemed and worked for the event, calls it foresight. But the actual average of events, which depend entirely on chance, will be the same.

If a man plays at *rouge et noir* with one chance in a hundred in favour of the bank, it is certain that if he plays often enough, he will lose his capital once at least for every hundred times he plays. Or, if he speculates on the Stock Exchange, the turn of the market and broker's commission will, in the long run, certainly swallow up his original capital. And yet men will gamble and speculate, because they cannot resist the pleasing illusion that they are lucky, and that it would be very nice to win a large stake without having had to work for it.

There is nothing for which self-knowledge is more indispensable in practical life than to enable a man to steer a straight course between opposite extremes, and to discern clearly the boundary line between right and wrong. The law of polarity, by which things good in themselves if pushed to extremes become bad, and every truth develops a corresponding error, is of daily and universal application in practical affairs.

Take, for instance, the much-debated question of the pursuit of money. Poets and novelists are never tired of denouncing the "Auri sacra fames," and there is no doubt that, when carried to excess, it is the fertile source of crime; and even in a less degree, it leads to meanness and dishonesty, and has a degrading influence on the individual or the nation who give themselves up too exclusively to the worship of the "almighty dollar." But, on the other hand, the desire, or rather the necessity under the conditions of civilised society, of making money, is by far the most powerful and all-pervading influence of practical life. And, within due bounds and under proper conditions, it is a healthy and beneficial influence. At the lowest stage it obliges men to work instead of being idle, and this is an immense advantage both to the community and to the individual. An idle man, in every grade of society, is generally a worthless and often a bad man; while an honest working man, whether the work be of the head or hand, is far more likely to be happy and respectable.

Again, the necessity of earning money is a wonderful test of the real value of a man in the world's market. We should be all very apt to become pretentious windbags of conceit, if we were not brought to our senses by the wholesome *douche* of having to work for a livelihood. Many a man who fancies himself intended for a poet or politician, and some who by accident of birth or fortune are pitchforked into prominent positions, would find it difficult to point out any occupation in which they are honestly worth a couple of hundred a year.

Even in the higher departments of art and literature, it may be questioned whether the healthy, natural desire to turn an honest penny has not inspired greater works than a morbid appetite for fame. Shakespeare's ambition was to retire to his native town with a moderate competency; Walter Scott's to become a laird, with a family estate, in the border-land of the chief of his clan—"the bold Buccleugh." And, in the present day, literature is becoming more and more an honourable profession, which men take to, as they do to law or medicine, as a means of earning a livelihood.

It must always be borne in mind that under the practical conditions of modern civilisation, money means not only the possibility of bare existence, but nearly all that makes existence tolerable—health, recreation, culture, and independence. The number and locality of the rooms a man lives in, the number of cubic feet and purity of the air he and his family breathe, are questions of rent; the food they eat, the clothes they wear, the books they read, the holidays they enjoy, are all questions of money. And above all, without money there is no independence. An absolutely penniless man has to fall back on crime or the workhouse; a poor man is at the mercy of a thousand accidents; sickness, fluctuations of trade, caprice of employers, pressure of creditors, may at any moment reduce him and those who depend on him to want. It admits of no question, that the first duty of every one is to endeavour to raise himself above this level of ignoble daily cares, and plant himself in a position where he can face the present and look forward to the future with tolerable equanimity.

As we rise in the scale of society the problem becomes more difficult. Money-making is very apt to be pushed to excess and lead to gambling and dishonesty; while the worship of wealth, which is perhaps the besetting sin of the age, is distinctly the cause of much lax morality and snobbish vulgarity. But on the other hand, money is power, and a large fortune honestly acquired and well spent, gives its possessor unrivalled opportunities for doing good. He can assist charities, patronise art, and if gifted with force of character and fair abilities may become a legislator and statesman, and enrol his name in the annals of his country. It is hard to say that if a man has an opportunity of making a large fortune honestly, and feels that he has it in him to use it nobly, he should refrain from doing so because moralists cry "Sour grapes," and tell him that riches are deceitful.

But for nothing is "self-knowledge" more requisite than to enable a man to see clearly how high he can safely aim, and what sort of stake he can prudently play for. The immense majority of mankind have neither the opportunities nor the faculties for playing for very high stakes, and must be contented with the safe game for moderate and attainable ends. One such end is within the reach of almost every one:

> To make a happy household clime
> For weans and wife,
> Is the true pathos and sublime
> Of human life.

So says Burns, who has a rare faculty of hitting the right nail on the head; and the ideal he sets before us in these simple lines is at once the truest and the most universal. The man who fails in this is himself a failure ; while the man who by his industry and energy supports a family in comfort and respectability according to their station, and who, at the same time, by control of temper, kindness, unselfishness, and sweet reasonableness makes his household a happy one, may feel, even though fortune may not have placed him in a position of higher responsibilities, that he has not lived in vain, that he has performed the first duties and tasted the truest pleasures of mortal existence, and that, whatever there may be behind the impenetrable veil, he can face it with head erect, as one of "Nature's gentlemen."

SELF-CONTROL.

This is, after all, the vitally important element of a happy and successful life. The compass may point truly to the pole, the chart may show the right channel amidst shoals and rocks, but the ship will hardly arrive safely in port unless the helmsman stands at his post in all weathers, ready to meet any sheer of the bow by a timely turn to starboard or to port. So self-reverence and self-knowledge may point out ever so clearly the path of duty, unless self-control is constantly present we shall surely stray from it. At every moment of our lives natural instinct tells us to do one thing, while reason and conscience tell us to do another. It is by an effort that we get up in the morning and go about our daily work. It is by an effort that we refrain from indulgences and forego pleasures, control our passions, restrain our tempers. The uncultured man is violent, selfish, childish ; it is only by the inherited or acquired practice of self-control that he is transformed into the civilised man—courteous, considerate, sensible, and reliable.

The necessity of self-control in all the more important relations of moral and practical life is so obvious that it would be only repeating commonplaces to enlarge on it. But there is often danger of its being overlooked in those minor morals of conduct which make up the greater part of life, and determine the happiness or misery of oneself and others.

For instance, control over the temper. A man never shows his cousinship to the ape so much as when he is in a passion. The manifestations are so exactly similar—irrational violence, nervous agitation, total loss of head, and abdication of all presence of mind and reasoning power. To see a grown-up man reduced to the level of a spoiled child, or of a monkey who has been disappointed of a nut, is a spectacle of which it is hard to say whether it is more ridiculous or painful. Even worse than occasional violence is the habitual ill-temper which makes life miserable to those who are obliged to put up with it. We call a man who strikes a woman or child with his fist a brute; what is he if he strikes them daily and hourly, ten times more cruelly, with his tongue? A ten times greater brute. And yet there are men, calling themselves gentlemen, who do this, either from sheer brutality of nature, or oftener from inconsiderateness, coarseness of fibre, and inability to exercise self-control in minor matters.

There is one very common mistake made, that of considering relationship an excuse for rudeness. The members of a family may relax something of the stiffness of company manners among themselves, but they should never forget that it is just as much ill-breeding to say a rude thing to a wife, a sister, or a brother, as it would be to say it to any other lady or gentleman. In fact, it is worse, for the other lady can treat you with contempt and keep out of your way, while the poor woman who is tied to you feels it keenly, and has no means of escape from it. Good manners are, in practical life, a great part of good morals; and there is something to be said for religions which, like the Chinese, lay down rules of politeness, and make salvation depend very much on the observance of rites and ceremonies intended to ensure courtesy and decorum in the intercourse of all classes of the community in daily life.

Although not so bad as the indulgence of a violent or morose temper, a great deal of unhappiness is caused by a fussy and fidgety disposition, which makes mountains out of molehills, and keeps every one in hot water about trifles. This is one of the common faults of idleness, as genuine work both strengthens the fibre to resist and leaves no time to brood over petty troubles.

The excuse one commonly hears from those who give way to these petty infirmities is, "that they cannot help it, they are born with thin skins and excitable tempers." This is the excuse of sloth and weakness. If, as the poet says,

> Man is man, and master of his fate,

what sort of an unmanly creature must he be who cannot master even the slightest impulse or resist the slightest temptation, and allows himself to be ruffled into a storm by every passing breath, like a shallow roadside puddle? If he will not try he certainly will not learn; but if he will honestly try to correct faults, he will find it easier every time, until the fancied impossibilities fade away and are forgotten. A man who is so much afraid of tumbling off that he will never mount a horse, may fancy that Nature has disqualified him for riding; but for all that, nine men out of ten, if obliged to try—say as recruits in a cavalry regiment—though they may not all turn out accomplished horsemen, will all learn to ride well enough for practical purposes.

It is peculiarly important for the young to set resolutely about correcting bad habits and forming good ones, while the faculties are fresh and the brain supple; for, in obedience to the law by which molecular motions travel by preference along beaten paths, every year cuts deeper the channels of thought and feeling, whether for good or evil. A brain trained to respond to calls of duty soon does so with ease and elasticity, just as the muscles of the blacksmith's arm or of the ballet-dancer's leg acquire strength and vigour by exercise; while, on the other hand, motion is a pain and self-control an effort to the soft and flabby limb or brain which has been weakened by self-indulgence.

It is scarcely necessary to say that for success in practical life, self-control is the one thing most needful. To take the simplest case, that of a young working man beginning life with health, knowledge of a trade, or even without it with good thews and sinews, he is the most free and independent of mortals, on one condition—that he has saved £10. With this, he is a free agent in disposing of his labour, he can make his contract with an employer on equal terms, he can carry his goods to the best market, and is practically a citizen of the world, ready to start for San Francisco or Melbourne if he thinks he can better himself. Without it, he is a serf tied to the soil, he cannot move from place to place, he must take whatever wages are offered him or starve.

But how to save the £10 ? That is a question of daily and weekly recurrence ; whether to spend an extra shilling in the pleasant way of going to a public-house and sitting with a pipe and a jug of ale by the fireside among jolly companions, or to forego the pleasure and save the shilling. A shilling a week saved will, in four years, give him the £10, and go a good way to establish habits which, if he is enterprising and goes to a colony, or clever and has any luck at home, may readily make the ten a hundred, or even a thousand pounds. So in every class of life, the man who gets on is the man who has schooled himself never to ask whether a thing is pleasant, but whether it is right and reasonable ; and who always keeps a bright look-out ahead, and does his best at the task, whatever it may be, that is set before him.

Education really resolves itself very much into teaching the young to acquire this indispensable faculty of self-control. The amount of positive knowledge, useful in after life, acquired at our English public schools, is really very little beyond the three R's. A boy who could teach himself French or German in five months spends five years over Latin and Greek, and in nine cases out of ten forgets them as soon as he leaves school or college. Almost everything we know that is worth knowing we teach ourselves in after life. But the discipline of school is invaluable in teaching the lesson of self-control. Almost every hour of the day a boy at school has to do things that are disagreeable and abstain from doing things that nature prompts, under pain of getting a caning from the master or a thrashing from other boys. The memory also is exercised, and the faculty of fixing the mind on work is developed, by useless almost as well as by useful studies. In this point of view even that *ne plus ultra* of technical pedantry, the Latin grammar, with its "Propria quæ maribus" and "As in presenti," may have its use in teaching a boy that no matter how absurd or repulsive a task may be, he has got to tackle to it or worse will befall him.

But it is in a moral sense that the influence of a good school is most valuable. The average boy learns that he must not tell lies, he must not be a sneak or a coward, he must take punishment bravely, and conform to the schoolmaster's standard of discipline and the school-boy's standard of honour. In this way the first lesson of life, stoicism, becomes with most English lads a sort of instinct or second nature.

For stoicism, after all, is the foundation and primary element of all useful and honourable life. Whether as Carlyle's "Everlasting No," or as George Eliot's advice to take the pains and mishaps of life without resorting to moral opium, the conclusion of all the greatest minds is that a man must have something of the Red Indian in him and be able to suffer silently, and burn his own smoke, if he is to be worth anything. And still more a woman, who has to bear with and make the best of a thousand petty annoyances without complaint. Men can bear on great occasions, but in the innumerable petty trials of life women as a rule show more self-control and moral fortitude. What would the life of a woman be who could not stand being bored with a smiling face, put up with the worries of children and servants with cheerful fortitude, and turn away an angry word by a soft answer?

There is much more that might be said, but my object is not to preach or moralise, but simply to record a few of the practical rules and reflections which have impressed themselves on me in the course of a long and busy life. I do so in the hope that perchance they may awaken useful thoughts in some, especially of the younger readers, who may happen to glance over these pages. This much I may say for them, I have tried them and found them work well. I have lived for more than the Scriptural span of threescore and ten years, a life of varied fortunes and many experiences. I may say, in the words which my favourite poet, Tennyson, puts into the mouth of Ulysses:

> For ever roaming with a hungry heart,
> Much have I seen and known, cities of men,
> And councils, climates, governments.

And the conclusion I come to is, not that of the Preacher, "Vanity of vanities, all is vanity," but rather that life, with all its drawbacks, is worth living; and that to have been born in a civilised country in the nineteenth century is a boon for which a man can never be sufficiently thankful. Some may find it otherwise from no fault of their own; more by their own fault;

but the majority of men and women may lead useful, honourable, and on the whole fairly happy lives, if they will act on the maxim which I have always endeavoured, however imperfectly, to follow—

FEAR NOTHING; MAKE THE BEST OF EVERYTHING.

SUPPLEMENTAL CHAPTER.

Gladstone's "Dawn of Creation" and "Proem to Genesis."
Drummond's "Natural Law in the Spiritual World."

SINCE the above work was written, two essays have appeared which require notice; one, from the celebrity of its author, the other from its extensive circulation. I refer to Mr. Gladstone's articles in the *Nineteenth Century*, on the "Dawn of Creation and of Worship," and on the "Proem to Genesis;" and to Professor Drummond's "Natural Law in the Spiritual World."

The first essay attempts to prove the inspiration of the Bible from the anticipations of the conclusions of modern science alleged to be contained in the Book of Genesis. The second, that of Professor Drummond, assuming this inspiration and the Calvinistic creed of theology based upon it, professes to show that the latter is the inevitable result of the same identical natural laws as prevail throughout the domain of Science.

I propose to deal first with Mr. Gladstone as a theologian.

"THE DAWN OF CREATION AND OF WORSHIP."

Mr. Gladstone's article in the *Nineteenth Century* on the "Dawn of Creation and of Worship" is exactly what might have been expected from him—eloquent, rhetorical, diffuse; anything, in short, except logical and closely-reasoned. His mental attitude towards these questions may be described in two words, as that of a man who is ecclesiastically-minded and Homerically-minded.

In fact, about one-third of his essay is taken up by a digression, which is almost entirely irrelevant, as to the extent to which the Olympian gods, as described by Homer, do or do not bear traces of being personifications of natural powers, and do or do not possess attributes which point to derivation from sources common to the author of the "Iliad" and the author of Genesis. It is needless to point out what a very remote bearing this speculation can have on the serious and vitally-important question, whether the account of the creation of the world and of man contained in the Bible is or is not consistent with the ascertained facts of modern science. That the Homeric gods are to a certain extent derived from solar myths is beyond doubt. Phœbus, the shining one, whose arrow-rays, darted in wrath, bring pestilence, is clearly in some senses the sun; and it admits of no question that the labours of Hercules are principally, if not wholly, taken from the signs of the Zodiac. But there are other elements mixed up with these, and if it should be proved that some of them are borrowed from ancient mythologies common to the Aryan and Semitic races, which is far from being an ascertained fact, it would go a very little way towards settling the question whether the narrative of Noah's ark is a true narrative.

The digression is chiefly interesting as illustrating the working of Mr. Gladstone's mind, which is eminently excursive, prone to elaborate details and to dwell on irrelevant issues to an extent which obscures the main argument. It is also a mind eminently sentimental and emotional, and he seems to think that questions of pure scientific fact can be decided by impassioned appeals to the feelings connected with old forms of faith. In such appeals it is needless to say that Mr. Gladstone is at home, and that those who are already convinced will find in this, as in his other writings, strains of lofty, if somewhat vague and verbose, eloquence to read and to admire. Nor can it be denied that any candid reader, whether convinced or not, must feel his admiration increased for a man who, amidst the exciting occupations of political life, can keep his mind open to such subjects and snatch a leisure hour to write upon them.

But when we pass from these side issues to the central question, we cannot allow our admiration for Mr. Gladstone to give more weight to his assertions and arguments than if they proceeded from some unknown Mr. Smith or Mr. Jones. The issue is quite definite and precise. Is or is not the account of the creation contained in the Old Testament *true*—that is, consistent with real facts which no one can dispute? Mr. Gladstone undertakes to prove that it is *true*, and that its accordance with facts, as ascertained by modern science, goes a long way to prove the inspiration of the volume in which it is contained.

To sustain this weighty proposition it is obvious that

the first requisite is to be thoroughly acquainted with the most recent discoveries in astronomy, geology, zoology, physiology, and, in fact, with all branches of modern science. The time is long past when the *facts* had to be tested by their correspondence with the *theory* of an inspired revelation; nowadays it is the *theory* which has to be tested by its correspondence with the *facts*. Mr. Gladstone enters upon this arduous contest with the gallantry and confidence of an Arab who takes the field armed with sword and spear, to oppose, for the first time, an adversary armed with rifle and revolver. He says himself that he is "wholly destitute of that kind of knowledge which carries authority," and the most cursory perusal of his essay is sufficient to show it. For instance, he states that the fourfold division of animated creation set forth in Genesis, viz.:

1. The water population;
2. The air population;
3. The land population of animals;
4. The land population consummated in man—

"is understood to have been so approved in our time by natural science, that it may be taken as a demonstrated conclusion and established fact." Is it possible that Mr. Gladstone never heard of the iguanodon of the Wealden, or of the small insectivorous and marsupial animals of the Oolite, or of the labyrinthodon and large batrachians of the Trias, or of the scorpion of the Silurian, all of which lived on land many millions of years before a single species of any fish now inhabiting the waters, or of any birds now inhabiting the air, had come into existence? Can he ever have visited the South Kensington Museum, and seen the fossil from Œuingen, of the feathered creature, half bird, half reptile? And is he ignorant of the great mass of evidence tending to show how the existing forms of bird life were developed from reptilian life, at a period enormously remote, but still long subsequent to the existence of many species of that "land population" which he complacently assumes that modern science has proved to have had no existence prior to the creation of the population of air and water? If Mr. Gladstone will go to the British Museum, he will see there a slab of sandstone from one of the very oldest formations, and probably deposited more than 100,000,000 years ago; and what will he see on this slab? Little pits made by rain-drops, higher on one side than the other, showing that the shower fell during a smart breeze; ripple-marks made by the tide exactly similar to those now made in the estuary of the Mersey or Solway, and numerous castings and tracings made in the wet sand by worms. What does this prove? That at this remote period the winds blew, the rain fell, the tides ebbed and flowed, implying the existence of their cause—the sun and moon; that an animal creation existed, which, as it lived entirely on land, although moist land, can hardly be described as falling within the category of either a water or an air population.

It would be easy to multiply instances, but it is superfluous to do so, when the late President of the Royal Society, Professor Huxley, the highest living authority on these questions, has so recently as in the last number of the *Nineteenth Century* said, "If I know anything at all about the results attained by the Natural Science of our time, it is 'a demonstrated conclusion and established fact' that the fourfold order given by Mr. Gladstone is not that in which the evidence at our disposal tends to show that the water, air, and land populations of the globe have made their appearance." To those who have the most elementary acquaintance with works like those of Lyell, Huxley, and Haeckel, the assumption that such a succession is proved by science must appear as amazing as if Mr. Gladstone had stated it to be a demonstrated conclusion that the earth was flat and not round. His other arguments in support of the Genesis account of creation are of the same nature: those of a man fifty years behind his time in everything that relates to modern science.

The history of creation contained in the first chapter of Genesis, if the words are taken in their obvious and natural meaning, is perfectly clear and consistent. It is, as Mr. Gladstone says, "a singularly vivid, forcible, and effective popular narrative; or, if we like to take it so, a sublime poem"—of what? Of the cosmogony common to the early thinkers of the ancient world, and which must inevitably have been the first conception of those who, in the infancy of science, began to attempt an explanation of the origin of the phenomena presented to the natural senses. Man and his habitation the earth were assumed to be the central and primary facts of the universe. The earth was first formed out of chaos; light separated from darkness, the seas from the land; and the whole surrounded by a firmament or crystal vault, solid enough to separate the waters above, which caused the rain, from the waters below, and to support the heavenly bodies which revolved with it in twenty-four hours round the earth. In this firmament the sun was placed to rule the day, and the moon to rule the night, and, as its name, "the measurer," denotes, to measure times and seasons. The stars also were added as things of minor importance, probably for ornament, or to aid the work of the moon in nights when the lunar orb was invisible. The inorganic world being thus created, the earth was conceived to have been peopled, once for all, with its existing animal life by three successive stages, viz., the fish, or water population; the birds, or air population; and land animals; and the whole work crowned by the creation of man in God's "own image." This work was conceived to have been carried out by an anthropomorphic Deity, or magnified man, who worked like a man, by regular spells of day-work, surveying each evening the work of the preceding day to see that it was properly done, and resting on the seventh day after his week's labours. This is the plain, simple, and obvious meaning which the narrative must have conveyed to every one to whom it was addressed at the time, as it did to every one who read it until quite recently. The question is, is it a *true* narrative, that is, consistent with the *facts* now established; and, if untrue, can the volume be inspired which contains mistakes on matters of such importance?

The first observation is, that to bring the question at all within the limits of reasonable discussion it is necessary to assume that the words of the narrative are to be taken in a non-natural sense; that is, in a sense different from the obvious meaning which they must have conveyed to those to whom they were addressed. This presents no difficulty to Mr. Gladstone, whose mind has a singular capacity for using words in this non-natural sense, and saying things which may mean almost anything that the different political or other proclivities of different hearers may choose to find in them. Thus he has no difficulty in assuming that the "firmament," which supports the stars and separates the waters, may mean simply an expanse; or that if the writer of Genesis says "days" he means "periods," notwithstanding their duration being expressly defined by an "evening and a morning;" and the reference to them as an authority for the seventh natural day being taken as a day of rest. It may be sufficient to say, that to ordinary minds such a use of language by any uninspired writer would be without hesitation termed "Jesuitical," and that there is absolutely no authority for it, except in the preconceived determination to escape, *per fas vel nefas*, from the too direct antagonism between Scripture and science. But waiving this point, and allowing the fullest latitude for non-natural meanings, the difficulty is only postponed. The assumption that Laplace's nebular hypothesis, or any other hypothesis at all consistent with known astronomical and geological facts, can in any way be reconciled with the "stages of the majestic process described in the Book of Genesis" is as untenable as that of a solid crystal vault, or of six literal days for creation. Mr. Gladstone argues that if the author of Genesis mentions the creation of the earth as the beginning of the work, and introduces the sun and moon only on the fourth day, he may have meant, not that the sun and moon had no previous existence, but that the "assignment to them of a certain place and orbit respectively, with a light-giving power," only took place long periods after the geological structure of the earth had been completed by the "emergence of our land and its separation from the sea." It is, of course, obvious that the first condensation of any cosmic nebula must have taken place about a central nucleus; in other words, about a sun, and that planets and satellites can only have been detached successively, and with their places and orbits assigned, as the rotating mass contracted. By no possibility could an intermediate planet like the earth have been detached out of its order before other members of its family.

Still more hazy are Mr. Gladstone's ideas respecting the separation of light from dark, and wet from dry. He seems to consider light and darkness as separate substances, which, like white and black beans mixed together in a bag, could be taken out and sorted into two separate heaps. No other sense can be attached to the employment of such a phrase as "the detachment and collection of light." It is, of course, well known that light is simply the vibration of an almost infinitely rare and elastic medium called ether, and darkness the absence of such vibration; and that cosmic matter, even in the earlier stages of nebulous formation, is self-luminous, *i.e.*, emits light. Light, therefore, must inevitably have long preceded the aggregation of this matter into the planet known as the earth. The "detachment of wet from dry, and of solid from liquid," is open to still more obvious objection. It is evidently the expression of one who supposed that the separation of sea from land was a process which took place, once for all, establishing the present configuration of the earth's surface, whereas it is certain that there has been a perpetual rising and sinking, and alternation of sea and land, going on from the earliest geological periods. The chalk, which now forms a large portion of continents and rises into considerable hills, was formed at the bottom of a deep ocean. The Wealden, which, below the chalk, is the Delta formation of a large river, implies the existence of a continent drained by that river which has long since disappeared beneath the chalk ocean.

And so on for all the stratified formations forming nine-tenths of the earth's crust, which must all have been formed beneath water by denudation of older rocks, and subsequently upheaved. Even in quite recent times, and since the appearance of man, Britain has been at one time an archipelago of islands in a frozen sea, and at others part of a continent, roamed over by the mammoth, the Irish elk, and the reindeer.

When we pass from inorganic to organic nature, the account of the creation of animated being is in still more direct opposition with facts. We have already seen what a mistake Mr. Gladstone commits in supposing that the succession of life was in the regular order of a water, an air, and a land population. But this is a mere nothing to the difficulty in reconciling the creation of those three orders of being in three successive days, with the enormous multitude of special miraculous creations required to account for the vast number of separate species actually existing in separate zoological provinces of the earth, and for the incalculably vaster number proved by their remains to have come into existence, flourished, and died out in the older geological formations. Madeira alone contains no less than one hundred and thirty-four species of land snails peculiar to this little group of islands, of which only twenty-one are found in Europe or Africa. If we discard the theory of evolution for that of miraculous creation, we must suppose the miraculous act to have been exerted one hundred and thirteen times in Madeira alone for no other purpose than that of giving it a variety of land snails.

It is, however, when we come to the creation of man that the discrepancy between the account in Genesis and

the discoveries of modern science strikes us most
forcibly. According to Genesis, "God created man in
his own image," at a date which, measured by years or
generations, is comparatively recent. In the time of
Cuvier, on whose authority Mr. Gladstone relies, no
geological evidence had been discovered to confute this
statement, and the supposed absence of human remains
in connection with extinct animals, or in anything older
than the merest superficial deposits, was reasonably
thought to give it considerable support. But the case
was completely altered when hundreds of thousands of
undoubted human remains came to be discovered in the
gravels of ancient rivers, and securely sealed under beds
of stalagmite in caves, associated with remains of extinct
animals, and under conditions implying enormous
antiquity. No one who has the slightest acquaintance
with the subject any longer doubts that Palæolithic man
must have existed at any rate during part of the Glacial
period, and in all probability much earlier. His existence on earth must be measured, not by generations
or centuries, but by long periods, the units of which
cannot be less than ten thousand years. It is equally
certain that these primeval men existed in a state of the
rudest savagery, and that, instead of falling from a high
state, the course of the human race has been that of
slow and painful progress upwards from rude and almost
bestial beginnings. These discoveries, of which not
even a hint escapes from Mr. Gladstone to show that he
is aware of them, have practically revolutionised the
attitude of modern thought towards old creeds. It is
no longer possible to consider as inspired revelations,
writings which contain views as to man's origin as diametrically opposed to actual facts as the legend of

Deucalion and Pyrrha, and very much farther from the truth than the account given in the poem of Lucretius.

If it requires some slight acquaintance with modern science to recognise fully the impossibilities involved in the account of creation given in the first chapter of Genesis, none is required to perceive the manifest impossibilities of what may be called the second creation of animated life, described in the narrative of the Noachian deluge. Mr. Gladstone makes no reference whatever to this, but it is as integral a part of the Bible as the account of the original creation.

What does this narrative tell us?

That God, seeing the wickedness of man, repented of having created him and the other inhabitants of the earth, and determined to destroy them; but that Noah, the one just man, found grace in His sight, and was warned to construct an ark, or big ship, in which to save from the impending flood himself and family, and a pair, male and female, of every living thing of all flesh, animals, birds, and reptiles. Another version makes the number of each species taken into the ark seven of each sex of clean animals and of birds, *i.e.*, fourteen instead of two; but the smaller number may be taken, so as to avoid the appearance of wishing to exaggerate the impossibility of the narrative. This being done, the flood came, and covered " all the high hills that were under the whole heaven," utterly destroying every living thing upon the earth, except those who were saved with Noah in the ark. The flood began on the 17th day of the second month—say the 17th February—and lasted at its height for a hundred and fifty days, the ark grounding on Ararat on the 17th July, and the tops of the other mountains being first seen on the 1st October. The ark was opened, and the animals came forth on the 27th February of the succeeding year, so that they were shut up rather more than twelve months. The account of Noah offering a burnt offering of every clean beast and fowl may be omitted, though clearly inconsistent with the first narrative, which says that only one male and one female of each species were preserved; nor is it necessary to dwell on the very rude anthropomorphic conception of God which represents Him as promising never again to destroy the earth because He was pleased by the sweet savour of the roast meat.

Compare this narrative with actual facts. In the first place, the number of cubic feet in an ark of the given dimensions is easily calculated, and it is apparent that it would be totally insufficient to accommodate pairs of all the larger animals, such as elephants, giraffes, rhinoceroses, bisons, buffaloes, oxen of various species, horses, asses, zebras, quaggas, elks, and the various species of the deer family, elands and other large antelopes, lions, tigers, bears, and other carnivora, to say nothing of all the enormous minor population of the earth, the land birds, reptiles, snails, insects, and so forth, which were all destroyed by a universal deluge flooding the whole earth for a year. To say nothing, also, of the vast stores of provender for the herbivora, and flesh for the carnivora, which must have been provided in the ark for more than twelve months' consumption, and of the impossibility of arctic and tropical animals living together for a year at the same temperature. Nor is the difficulty less, when they emerged from the ark, of seeing how the herbivora could exist until a new vegetation had sprung up on the earth soaked and sodden by being for a year under water, or how the carnivora could exist without preying on the single pairs of herbivorous animals, which were the sole tenants of that earth for long afterwards. Nor is it possible to account for the actual distribution of animal life in different geological provinces if it all radiated from the common centre of a mountain in Armenia. Could the kangaroo, for instance, have jumped at one bound from the top of Ararat to Australia, leaving no trace of its passage in any intermediate district? Or how can the narrative be reconciled with the fact of the existence, long prior to any possible date of the Noachian deluge, of an enormous variety, both of species and types of land life, which were gradually developed into more and more specialised forms, and which appeared at different periods, grew, flourished, and finally decayed and disappeared? Was the mammoth, whose skeleton, still covered with flesh and hair, was discovered on the frozen banks of the Lena, a descendant of a pair of mammoths who were saved in the ark; or the Elephas meridionalis, whose bones, twice the size of the largest existing elephant, are found in the forest bed at Cromer; or the anthropoid ape and sabre-toothed tiger of the Miocene; or the palæotherium and anoplotherium of the Eocene, or any of the earlier inhabitants of the earth's land surface?

No stretching of days into periods, or other use of words in a non-natural sense, can in the slightest degree get over the glaring contradiction between the *naïve* and almost infantile story of Noah's ark, and the facts, I will not say of science, but of common sense and common observation, which are patent to every decently well-read schoolboy of the rising generation.

The real "dawn of creation" is that traced through three different lines of scientific research:

First, that of astronomy, showing the progressive condensation of nebulæ, nebulous stars, and suns in various stages of their life history.

Secondly, that of geology, commencing with the earliest known fossil, the Eozoon Canadiense of the Laurentian, and continued in a chain, every link of which is firmly welded, through the Silurian, with its abundance of molluscous, crustacean, and vermiform life, and first indication of fishes; the Devonian, with its predominance of fish and first appearance of reptiles; the Mesozoic, with its batrachians; the Secondary formations, in which reptiles of the sea, land, and air preponderated, and the first humble forms of vertebrate land animals began to appear; and finally the Tertiary, in which mammalian life has become abundant, and type succeeding to type and species to species, are gradually differentiated and specialised, through the Eocene, Miocene, and Pliocene periods, until we arrive at the Glacial and Prehistoric periods, and at positive proof of the existence of man.

Thirdly, the line of embryology, or development of every individual life, from the primitive speck of protoplasm, and the nucleated cell in which all life originates, passing, as in the parallel case of types and species, through progressive stages of specialisation from the lowest, the amœba, to the highest, man—who, like all other animals, originates in a cell, and is developed through stages undistinguishable from those of fish, reptile, and mammal, until the cell finally attains the highly specialised development of the quadrumanous, and, last of all, of the human type.

In like manner the "dawn of worship" is to be found in the flint hatchets and other rude implements deposited with the dead, as by modern savages, testifying to some sort of belief in spirits and in a future existence. This clearly prevailed in the Neolithic, and possibly in the immensely older Palæolithic period, though the evidence for the latter is at present very weak, and the first object which can be affirmed with any certainty to be an idol or attempt to represent a deity, dates only from the Neolithic period, as do the cannibal feasts, which can be proved to have not infrequently accompanied the interment of important chiefs. For anything beyond this we have to descend to the Historical period, and turn to early monuments, myths, and sacred books. The earliest records by far are those of the Egyptian tombs of the first four dynasties, and they tell us little more than this, that with a highly developed civilisation, the idea of a future life was very much that of a continuance of the present life, in a tomb which was made to resemble the deceased's actual house, and with surroundings which repeated his actual belongings; while the whole complicated Egyptian mythology, of symbolised gods and deified animals, was of later origin. If we turn to the earliest mythologies of the Aryan and of the mixed Semitic and other races of Western Asia, we find them plainly originating, to a great extent, in the personification of natural forces, mainly of the sun, on which are engrafted ideas of family, tribal, and national gods, and of deified heroes. Sometimes, as the original meaning of the names and attributes of these gods came to be forgotten, the mythologies branched out into innumerable fables; at other times, among more simple and severe races, or with more philosophic minds in the inner circle of a hereditary priesthood, the fables of polytheism were rejected, and the idea prevailed, either of a unity of nature implying a single author, or of such a preponderance of the national God over all others as led by a different path to the same result of monotheism. The real merit of the Jewish race and of the Hebrew scriptures is to have conceived this idea earlier, and retained it more firmly, than any of the less philosophical and more immoral religions of the ancient world; and this is a merit of which they can never be deprived, however much the literal accuracy, and consequently the inspiration and miraculous attributes, of these venerable books may be disproved and disappear.

Works like this of Mr. Gladstone's, however well intentioned, are in reality profoundly irreligious, for if—like the throw of the gambler, who, when the cards or dice go against him, stakes all or nothing on some desperate cast—religion is staked on the one issue that incredible narratives are true, and were dictated by Divine inspiration, there can be but one result. Every day brings to light fresh discoveries confirming the conclusions of science, and conflicting with the accounts of the creation of the world and man, and of the universal deluge, given in the Old Testament. Every day diffuses a knowledge of these discoveries more widely among millions of readers. What must be the result if men of "light and leading" proclaim to the world that if these conclusions of science are true there is an end of religion? Evidently the same as George Stephenson predicted for the cow who should stand on the rails and try to stop the locomotive, "Varra awkward for the coo." The really religious writers of the present day are those who, thoroughly understanding and recognising the facts of science, boldly throw overboard whatever conflicts with them, abandon all theories of inspiration and miraculous interferences with the order of nature, and appeal, in support of religion, to the essential beauty and truth in Christianity underlying the myths and dogmas which have grown up about it; who, above all, appeal to the fact that it exists and is a product of the evolution of the human mind, satisfying, as nothing else can do so well, many of the purest emotions and loftiest aspirations, which are equally a necessary and inevitable product of that evolution.

" PROEM TO GENESIS."

Mr. Gladstone's first essay having elicited a crushing and conclusive reply from Professor Huxley, he followed it up by a second one under the above title, which is chiefly remarkable for the rhetorical dexterity with which he withdraws under a cloud of smoke from the positions rendered untenable by the Professor's heavy artillery, while at the same time he defends the equally untenable positions, not within his opponent's line of fire, by reiterated assertion.

Professor Huxley shows that the real facts, as ascertained beyond all doubt by the researches of science, do not correspond with the order of animated creation described in Genesis. Mr. Gladstone admits that this "pulverises his proposition that there was a scientific *consensus* as to a sequence like that of Genesis, in the production of animal life as between fishes, birds, mammals, and man." He rides off by saying that the writer of the account of creation in Genesis "is not responsible for scientific precision, that nothing can be assigned to him but a statement general, which admits exceptions; popular, which aims mainly at producing moral impressions; summary, which cannot but be open to more or less of criticism in detail."

In a word, he says, "I think it is a sermon." But how is an account of creation evaporated into a sermon to prove revelation?

Partly by evaporating revelation, which he tells us does not require us to believe that the Bible is strictly and literally true in all its statements, but that it may have a human element of error and uncertainty in the sacred text. This is virtually giving up the whole case, for it opens the door for human reason to inquire at every step, by the ordinary rules of criticism, whether any given statement is part of the divine inspiration, or part of the " human element " of error.

Mr. Gladstone sees this, and shrinks from carrying out this line of reasoning to its legitimate conclusions. Accordingly he falls back on so much of his original assertion as has been left undemolished by Huxley, and endeavours to prove it by repeating it. Admitting that "the statements of Genesis as to plants and reptiles cannot in all points be sustained," he contends that enough remains to prove revelation, notwithstanding these material errors, from the facts, "First, that such a record should have been made at all. Secondly, that instead of dwelling on generalities, it has placed itself under the severe conditions of a chronological order, reaching from the past *nisus* of chaotic matter to the consummated production of a fair and goodly, a furnished and peopled world. Thirdly, that its cosmogony seems, in the light of the nineteenth century, to draw more and more of countenance from the best natural philosophy;

and, Fourthly, that it has described the successive origins of the five great categories of present life with which human experience was and is conversant, in that order which geological authority confirms." The first point may be briefly dismissed. All religions, down to those of the rudest tribes, begin with cosmogonies. That of the Chaldees begins, like that of Genesis, with chaos, describes the separation of sea and land, and ends with the creation of man; only Bel is said to have made him from clay animated with his own blood, while Jahveh is said to have made him from the dust of the earth.

The gist of the question lies in the third and fourth propositions as tested by the second.

Now it is precisely this particularity of statement which brings the narrative of Genesis into contradiction with facts. If it had only stated that the universe of sun, moon, stars, and earth had been evolved from chaos, and that life had appeared on the earth in a gradation from the lowest to the highest types, culminating in the creation of man, there would have been nothing in it opposed to modern science, and, on the contrary, it might have been accepted as a wonderful anticipation of its discoveries.

But when it states, "under the severe conditions of a chronological order," that the earth was created on the third day, as defined by an evening and a morning; and the sun, moon, and stars on a subsequent day; and when the vault of heaven is described by a Hebrew word, which, while it expresses the idea of expansion, expresses also that of solidity, sufficient, as we are told both here and in the account of the deluge, to uphold waters and support the heavenly bodies—a sense which is given to it by all ancient authors and by the translation of the Septuagint—it becomes evident that the statement that "this cosmogony seems, in the light of the nineteenth century, to draw more and more of countenance from the best natural philosophy," is as amazing as that respecting the order of animated creation which has been "pulverised" by Professor Huxley.

How could a firmament, which was a mere expanse, support water, and let it down by opening "the windows of heaven," when rain was required for a universal deluge? And how could there be a "day" defined by an "evening and a morning," before the sun had had "assigned to it a certain place and orbit and light-giving power," and if it existed at all, existed only in the form of a diffused and non-luminous nebulous haze? Evening and morning are perfectly definite terms, which imply the existence of the sun and the opening and closing of a natural day of twenty-four hours, either by the apparent revolution of the sun round the earth, or by the real rotation of the earth round its axis, in either case in a "certain place and orbit" of the sun, and with a "light-giving power."

The only attempt to support Mr. Gladstone's original proposition is contained in the reiterated *assertion* that he is not aware that any serious flaw is alleged in the cosmogony of the Proem, and as regards its account of the creation of animated life in the argument that the words probably meant mammals only, and that the Mosaic writer only meant such animals "as were familiarly known to early man." But the words are most express; and the serpent, who belongs to the order of reptiles which existed before birds, and from which birds were probably developed, was certainly one of the land animals with which Eastern nations were most familiar, and with which men had the closest connection, as is shown by the narrative of the Garden of Eden, and the traces of Naga, or snake worship, which are found in so many primitive and rude religions.

But, after all, the enormous difference between the Biblical account of man's origin and *fact* comes out most clearly in the narrative of Adam's fall and of the Noachian deluge.

The impossibilities of the latter have been clearly pointed out; and it only remains to add that it requires us to believe that all the existing races of mankind—Aryan, Semitic, Mongol, Malay, Negro, Negrito, Papuan, American, Australian, and a multitude of others—have been developed from one family in less than 4,000 years, while we know, as a positive fact, from the Egyptian temples, that the most marked of these races, the Negro, existed, with all its present characteristics, more than 5,000 years ago, and has not varied perceptibly during that period.

As regards Adam's fall, the discovery of Palæolithic man is that which has really given the greatest shock to received theological opinions; for this discovery, which is an entirely new one of the last half century, though now confirmed by innumerable instances, not only flatly contradicts the narratives of recent descent from Adam and Noah, but it assails, in its most vital point, the whole dogma of Pauline Christianity.

The two statements cannot both be true: one, that man has fallen, the other, that he has risen; one, that he was created in God's image, with high moral and religious faculties, and placed in a garden in a state of innocence and happiness, from which he fell by an act of disobedience, entailing a curse on his descendants only partially redeemed by the atonement; the other, that he is the product of an evolution, tending ever upwards, over immense geological periods, from savages who chipped their rude flints on the banks of frozen rivers, chased the mammoth and the reindeer on the plains of Southern France, and held their cannibal feasts in caves excavated by small streams which ran 100 feet above their present level.

Which is true? And can the book be inspired which gives a totally false account of such a vital matter? This is the real question, of which Mr. Gladstone's two eloquent essays scarcely even attempt to touch the outer fringe.

DRUMMOND'S "NATURAL LAW IN THE SPIRITUAL WORLD."

It is not surprising that this work has had an immense circulation. It professes to do exactly what multitudes of readers are anxious to see done, viz., to reconcile science and religion, and show that the dogmas of theology are not only not inconsistent with natural laws, but actually based upon and identical with them.

Professor Drummond brings to this task many qualifications. He enters the arena, not like the great majority of orthodox writers, armed only with the obsolete bows and arrows of theological infallibility, but equipped with the improved weapons of modern scientific research. He understands what is meant by laws of nature, and does not misrepresent or ignore them. He is learned, he is candid, and he is sincere. His style is clear, and his arguments and phraseology are such that, while the few who have scientific knowledge can understand and appreciate them, the many, who do not understand, cannot fail to find them profound and convincing where they chime in with their preconceived opinions.

It is the more necessary, therefore, for those whose sole object is that truth should prevail, to examine the work closely, and endeavour to show clearly what it aims to accomplish, how far it succeeds, and how far its conclusions are vitiated by underlying fallacies.

The fundamental idea of Professor Drummond's work is summed up in its title : " Natural Law in the Spiritual World."

The object is to prove that the same laws of nature which prevail throughout the organic and inorganic worlds of science extend, with an unshaken and identical continuity, into the world of spirits, and give positive and scientific proof of the dogmas of religion.

To establish this it is clear that the fundamental requisite is to begin with a precise definition and sufficient proof of the two terms of the proposition : " What is 'Natural Law'? What is the 'Spiritual World'?"

If, for instance, the proposition were that the same identical law of gravity prevails in the astronomical and geological worlds, we should have to begin by having a clear idea of what we mean by astronomy and what by geology. Astronomy means the knowledge of the sun and its planets, of satellites, comets, meteors, and those distant suns and systems called stars and nebulæ as far as their nature is disclosed to us by the telescope and spectroscope. The law of gravity is shown to prevail universally throughout this world, by experiments in the fall of heavy bodies, and calculations from the observed orbits of all heavenly bodies, from the solar system to the remotest double stars.

Geology, again, is the science of the formation of the planet which we inhabit, with its succession of strata upon strata, slowly deposited, frequently depressed and elevated, and identified by various types of life, appearing, growing, declining, and dying out, in the different formations. The prevalence of the identical law of gravity throughout the vast periods of time embraced by geology is easily proved from the phenomena of denudation and stratification. It is clear that heavy bodies have always gravitated, as they now gravitate, towards the earth's centre, and that, throughout those remote periods, mountains have been washed down by rivers into the sea—as they are now being washed down—and there subsided into stratified masses, which have been brought up again by repeated upheavals.

But to feel this certainty, we must have a clear idea of geology, and not a vague one, which may include all manner of catastrophes, miraculous interferences, and other phenomena unknown from any experience of existing nature.

Apply this to Professor Drummond's proposition.

Its first term is clear enough; there can be no doubt what he means by natural laws, and no one can define them more forcibly and distinctly.

He tells us: "No man can study modern science without a change coming over his view of truth. What impresses him about nature is its solidity. He is then standing upon actual things, among fixed laws."

And again:

"There is a sense of solidity about a law of nature

which belongs to nothing else in the world. Here at last, amid all that is shifting, is one thing that is sure; one thing outside ourselves, unbiassed, unprejudiced, uninfluenced by like or dislike, by doubt or fear."

But how of the other term of the proposition, the "Spiritual World"?

Here there is no attempt at definition, and even the fact of its existence is asserted and not proved.

In his Introduction, all he says about it is that "his proposal does not include an attempt to prove the existence of the spiritual world. *Does that need proof?*"

No, if you are content to keep to the sphere of theology, and accept authority or intuition as sufficient proofs. But yes, if you appeal to Cæsar, and asked to be tried by Cæsar's laws; in other words, if you attempt to prove religious dogmas by scientific reasonings.

He tells us, "The facts of the spiritual world are as real to thousands as the facts of the natural world." So were the facts of witchcraft and demonology. Does it prove them to be true? How is it possible to decide whether certain laws do, or do not, apply to the spiritual world, as long as we are left in entire uncertainty as to what may be meant by it, and whether it is intended to include everything that is not strictly matter, such as human consciousness, individuality, intellect, and morality; or to be confined to the particular tenets of one particular religious sect? How can we argue with a man about the laws of the spiritual world, without knowing whether he is a Plato, a Confucius, or a Comte, who embraces the whole sphere of humanity; or a Muggletonian or Plymouth Brother, whose idea of it is limited to the world of those who have been touched by Divine grace to believe in the special doctrines of his own minute congregation?

In the present instance Professor Drummond's conception of the "Spiritual World" is to be gathered, not from any precise definition, but from a careful perusal of his entire work, and by a gradual process of eliminating all that he affirms to form no part of it.

Thus, we gradually discover that *all* the natural elements of humanity, *all* that can be discovered by natural reason, *all* that can be explained and demonstrated, lie totally outside his idea of the "Spiritual World."

"What now," he says, "specifically distinguishes a Christian man from a non-Christian man? *Not* a higher morality, nobler character, benevolent sympathies, and reverent spirit. The distinction between them is the same as between the organic and inorganic, the living and the dead."

And again:

"Were we to construct a scientific classification, science would compel us to arrange all natural man, moral or immoral, educated or vulgar, as one family. But the spiritual man is removed from this family utterly, by the possession of an additional characteristic."

What is this characteristic?

The Professor asks and answers the question in the following terms:

"What is the something *extra* which constitutes spiritual life? It is Christ. He that hath the Son hath life."

He repeats this over and over again with ever-increasing emphasis.

"The earthly mind may be of noble calibre, enriched by culture, high-toned, virtuous, and pure. But if it know not God?"

"The Christian is an unique phenomenon. You cannot account for him. And if you could, he would not be a Christian."

If so, I am afraid the Professor's attempt to account for him by biogenesis and other natural laws, must be set down as endeavours to extinguish this "unique phenomenon," and banish Christianity from the world; for, if this definition be true, there would no longer be any Christians, if Christianity could be accounted for by rational arguments.

But it would be unfair to take advantage of a slip of the pen, or exaggeration of language, and we pass this over to inquire what, after these explanations, Professor Drummond's "Spiritual World" really amounts to.

It is evident that it is simply our old friend, the "Shorter Catechism," in a scientific dress. In other words, it is the world of Calvinistic Christianity—of the peculiar system of theology which turns on the ideas of original sin, fall, redemption, regeneration, election, and predestination.

The Professor does not shrink from setting forth this theory in all its grim repulsiveness.

"It is an old-fashioned theology," he says, "which divides the world in this way, which speaks of men as living and dead, lost and saved—a stern theology all but fallen into disuse. Nevertheless, the grim distinction must be retained. It is a scientific distinction. He that hath not the Son hath not life."

That is to say, that no amount of moral excellence

or intellectual superiority, ever has saved or ever can save the natural man from the curse of death, entailed on him by Adam's act of disobedience, and that a limited number of elect only can escape from it and inherit eternal life by virtue of the atonement, and "the breath of God *blowing where it listeth*, touching with its mystery of life the dead souls of men, and bearing them across the bridgeless gulf between the natural and the spiritual."

This proposition, he tells us, is, in the first place, made known to us and proved by revelation, and then confirmed by showing that it is the result of the same identical natural laws as those which prevail in the domain of science.

The law on which he mainly relies is that of biogenesis, which, he says, "is the fundamental law of life for both the natural and spiritual worlds."

Biogenesis means, that as far as is at present known, all life seems to originate from pre-existing life, and that the passage from the inorganic world of dead matter to the organic world of life, is only made in some unexplained way, which implies the intervention of some agency not reducible to known laws of science, and which therefore may be regarded as supernatural. From this he argues that the same supernatural agency must be assumed to continue throughout higher spheres of existence, and bridge the passage from the natural to the spiritual world just as it bridges that from atoms of carbon, oxygen, hydrogen, and nitrogen into protoplasm.

The first remark is that biogenesis is by no means a demonstrably certain and universal law like that of gravity. It simply amounts to this, that up to the present time no demonstration has been given that life can be produced otherwise than from pre-existing life ; and that certain experiments which appeared to establish the reality of spontaneous generation, have been shown to be fallacious. But the best scientific authorities who have been foremost in detecting the fallacy of these experiments, are also foremost in declaring that as a question of probability and not of positive proof, their belief is that at some earlier stage of the earth's existence, under conditions of heat, pressure, and electricity, different from those we can now produce in our laboratories, this passage from the inorganic to the organic has taken place, and no one would be greatly surprised to hear to-morrow of some experiment by which protoplasm had really been manufactured from chemical elements.

It is a long way from this to the certainty and universality of laws like those of gravity and the conservation of energy.

But waiving this objection, and supposing that biogenesis was really a certain law, what would it teach us? By whatever process we attempt to sound the depths of the universe, we soon arrive at the end of our tether, and are arrested by the Great Unknown, which we have no faculties enabling us to penetrate. From nebulæ to stars, from stars to suns, from suns to planets, from planets down through molecules to atoms, we can explore our way, and connect all phenomena by continuous laws. But what lies behind the atoms, what are they, how came they there? We know as little as we do of life, if there be life, in Saturn; or of what space may contain beyond the limits reached by the most powerful telescope.

If biogenesis really be a law, it simply brings us one step nearer to this Great Unknown by following the line of living matter up to protoplasm, than if we follow that of inorganic matter up to atoms. It is no more possible to prove theological dogmas from the laws of protoplasm than it is from the atomic theory.

All attempts to prove the extension of natural laws from one sphere into another which is not *in pari materia* with it, really resolve themselves into analogies, and no one is more aware than Professor Drummond of the danger in such cases of trusting to analogy.

He says: "The position we have been led to take up is not that spiritual laws are analogies to natural laws, but that they are the *same laws*."

And again:

"Nothing could be more false both to science and religion than attempts to adjust the two spheres by making out ingenious points of contact in detail."

The difference between analogies and proofs where the subject is not *in pari materia*, will be at once apparent if we consider the analogies between the world of nature and that of the human mind. Poetry consists, to a great extent, in the faculty of vividly conceiving and expressing such analogies. When Byron compares the flash of lightning in a midnight storm among the Alps, to the

> Light of a dark eye in woman,

it is beautiful poetry.

But does this prove that because the tumult of passion in a woman often ends, like the thunderstorm, in a shower of tears, therefore the same identical laws of electricity which cause the rain cause the tears?

This is Professor Drummond's proposition, and its fallacy will be at once apparent.

Again, the danger of founding religions on analogies will be apparent, if we consider what the consequences would be of extending this mode of reasoning to other religions than Christianity, and other laws than biogenesis.

There is no more certain or more universal law than that of the "conservation of energy," but if the human soul is not a mere attribute of matter, but an independent energy, it follows, if this law extends to it, that it can never die, but only be transformed. The Calvinistic theory of death for the immense majority, and life for the few elect, disappears, and instead of it we have a religion like that of the Brahmins and Buddhists, teaching the transmigration of souls from one form of life to another, and the final absorption of all the separate rills of individual life in the great ocean of Pantheism.

Again, polarity is a most universal law, and here we really do know that it extends not only throughout the inorganic and organic worlds, but also into what may be called the natural spiritual world—the world of laws and morals, of arts and sciences, of practical conduct, of social and political problems. There is scarcely a question to which the apologue does not apply of the knights who fought because they could each see only one side of the shield, and with reference to which true wisdom has not, in the words of the poet, to

> Turn to scorn with lips **divine**
> The falsehood of extremes.

What follows? Shall we embrace the religion of Zoroaster, which certainly gives us the best embodiment of this all-prevailing law, and presents it to us in the form best adapted for a great many of the realities of practical life?

The real truth is that religion, in the sense in which Professor Drummond uses the words "spiritual life," and in which the majority of the Christian world accept it, can only be proved by revelation. This he admits himself, for he says: "The revelation must be assumed. The information, in the first instance, must be vouchsafed as a revelation."

The truth therefore of any system of theology, which professes to teach things undiscoverable by ordinary human reason, must depend on two things.

Firstly, the evidence for the revelation by which it is made known.

Secondly, its accordance with other known and undoubted natural laws.

The second point may be considered first, for, although natural laws cannot of themselves discover or prove dogmas beyond the province of natural reason, yet, as all truth must be consistent with itself, it is not possible to believe in any revelation which professes to teach things absolutely irreconcilable with fundamental laws, either of the scientific or of the moral and intellectual worlds. No educated man could sincerely believe a theology which taught that the earth was flat and not round, that the law of gravity was that of the inverse cube and not the square of the distance, or that cruelty and ingratitude are virtues and not vices.

Tried by this test, the weakness of Professor Drummond's assumption of a "spiritual world," based on the lines of Calvinistic theology, is at once apparent.

Stripped of high-sounding theological language, and stated in plain English, what does it amount to?

Suppose we read in Herodotus a narrative how some great Asiatic king of kings—say, Cambyses, the son of Cyrus—offended by some act of disobedience on the part of the governor of a province, made a decree sentencing all the inhabitants of that province to be put to death on attaining a certain age; how the monarch's only son interceded for them with his father, but was told that the laws of the Medes and Persians could not be changed, and that none of the inhabitants of the province could escape the penalty unless the son offered himself as a sacrifice and atonement for them; how the son, being a noble and generous character, offered himself accordingly, and was put to a painful death; and thereupon the monarch remitted the penalty, not to all, but to a very small percentage of the inhabitants of the province, selected by lot, or by his favour, "blowing where it listed."

Will Professor Drummond, or any one else, tell us how this narrative differs from the Calvinistic scheme of theology, or how it can be reconciled with those moral laws of justice, mercy, and loving-kindness, which have come to be fundamental laws in the conceptions and consciences of all civilised races of mankind?

If it were possible to conceive of any revelation of such a scheme, supported by evidence so cogent and irresistible that it was impossible to doubt it, the only logical conclusion would be that the divine scheme of the universe was that of Zoroaster; a polarity between the two opposing principles of good and evil, the latter embodied in the father and the former in the son. But the practical conclusion would probably be either blank scepticism, or a belief that there was a mistake somewhere, either in the revelation, or in the interpretation of it.

This, at any rate, is clear, that the evidence for such a revelation must be of the most cogent and convincing character to induce any reasonable man, who approached the subject without prepossession, to entertain it for a moment, and that without such evidence no possible analogy between the scheme and some one or two out of the many laws of nature, could induce him to believe it.

Now, this is precisely the point which the defenders of orthodox theology ignore or overlook. The ever-increasing scepticism of the age, of which they complain, is based, not upon refined philosophical speculations, or abstruse arguments, but upon certain plain and matter-of-fact considerations, which the discoveries of modern science have forced on the minds of thinking men.

Orthodox Christianity is based on revelation ; what is revelation based on ? On the Bible—the whole fabric depends on the belief that the Bible is an inspired record conveying a Divine message from God to man.

Such a record it is clear cannot contain errors and contradictions upon material points affecting the whole scope and tenor of the message. If it appear, upon strong *primâ facie* evidence of scientific laws and facts, that it does contain such errors and contradictions, faith must be shaken; and the first condition of restoring it must be to show, either that these scientific facts are mistaken, or that the accounts in the Bible can be reconciled with them. Thus, for instance, the account of the creation of the material universe, earth, sun, moon, and stars, given in Genesis, seems to be absolutely inconsistent with the real facts as ascertained by astronomy and geology — that of the animated creation still more so, whether as described in Genesis, or even more palpably in the narrative of the deluge, and what may be called the second creation of life, in which all the varieties of the human species and the whole innumerable varieties of animal land life are said to be descendants of single pairs who were prisoned together in the ark for more than twelve months, and whose progeny radiated only some 4,000 years ago from a single centre on a mountain in Armenia.

And most destructive of all to old beliefs, the recent discoveries of the remains of Palæolithic man shatter into fragments the account of man's descent from an Adam created quite recently, in God's image, with high faculties, in a state of innocence and happiness, from which he fell by an act of disobedience; and again, still more palpably, from a Noah who was saved in an ark at a date not nearly so remote from us as the historical monuments of the earlier Egyptian dynasties.

If the facts really are that man's existence on the earth can be traced back for enormous periods, during which he has slowly but constantly progressed from a state of the rudest savagery towards civilisation and morality, how can this be reconciled with the theory of Adam's fall, which is the foundation of the whole superstructure of redemption and regeneration?

And how can the facts be denied, unless we are prepared to admit that the many hundreds of thousands of Palæolithic remains found, from Europe to China, have been placed there by a conspiracy of all the geologists of the world, to forge proofs contradictory of the Mosaic narrative?

Again, if we turn to the New Testament, is it possible to consider writings inspired which contain the most distinct and definite prophecy that a certain event, the end of the world, would take place within a certain definite period, the lifetime of some of the existing generation, when, in point of fact, it did not occur, and has not occurred, for nineteen centuries afterwards? Or, how can we believe them inspired, if some of the principal witnesses say of the cardinal miracle of the ascension, that they were commanded to go to Galilee to witness it, while others, who describe it fully and in detail, say that they were commanded not to go to Galilee, but to remain in Jerusalem, where the miracle actually took place? Or how can we account for the oldest manuscript of the Gospel, which is certainly one of the nearest, if not the nearest, to the original narrative, that according to St. Mark, omitting altogether any mention of any miraculous event connected with the resurrection?

These are the sort of difficulties which force themselves on the minds of all who have the most elementary acquaintance with the facts of modern science and the researches of Biblical criticism. They are plain questions which require a plain answer; and, until it is given, it is idle to appeal to authority and tradition, or to think that any amount of ecclesiastical scolding, or appeals to misty metaphysics, or far-fetched analogies to natural laws, can restore the simple creed of our ancestors, or prevent the faith of educated men from becoming every day fainter and fainter. If an answer can be given, by all means let it be given. Why should men, like Professor Drummond, who understand what natural laws really mean, and are conversant with the discoveries of modern science, be content to base the whole case for their "spiritual world" on texts from St. Paul and St. John, leaving the real foundations of belief to be defended by champions who rush into the field with the intrepidity of ignorance, and injure the cause they advocate by the obvious weakness of their arguments?

Let Professor Drummond, or any one else who is thoroughly acquainted with the latest discoveries of astronomy, geology, zoology, biology, palæontology, and Biblical criticism, face the real difficulties of orthodox belief, and show by reasonable arguments how science and religion can be reconciled, and he will meet with no prejudiced opposition. On the contrary, the great majority of mankind, including men of science, will be only too glad to be able to exchange the fainter for the larger hope. Because a man is acquainted with the facts of science he is not enamoured of annihilation, and would be delighted to find some secure basis on which to rest hopes of a future life, and of again seeing lost and loved faces; nor could he object to any additional sanction from revelation being given to the Sermon on the Mount, or St. Paul's definition of Christian charity.

But, if he is acquainted with these facts, and at all imbued with the spirit of scientific inquiry which characterises the age in which he lives, he asks for evidence; not absolutely certain or demonstrative evidence, like that for a proposition of Euclid, but reasonable evidence, such as, after standing being called into court and cross-examined, would satisfy a competent and impartial jury.

Such evidence has not hitherto been forthcoming, and assuredly is not supplied by this work of Professor Drummond's. In the meantime objections—not captious, but real, solid, reasonable objections—are multiplying every day; and every attempt to answer them makes it only clearer that the old theology rests on assertion and authority, and not on fact and argument.

When the point of attachment of a chain has given way it becomes almost a work of supererogation to test the strength of each separate link. It may be sufficient to say that, starting from the *assumption* that the spiritual world is identical with the Calvinistic creed, and that the truth of this creed is proved by revelation and confirmed by biogenesis, the rest of Professor Drummond's work consists of attempts to preach science into this creed, and show that its peculiar tenets have analogies in other natural laws.

Thus, the law of degeneration, by which organs dwindle and disappear by want of use, is used as an analogy for the decay of faculties by neglect, which, if cultivated, might have raised the soul to a higher level.

The law of growth is quoted in support of the doctrine of election, as showing that in either case the growth is not the result of conscious effort, but of some miraculous gift conferred by the grace of God "blowing where it listeth." We cannot, "by taking thought, add a cubit to our stature;" and by the same law we are told that we cannot—by any amount of conscious effort, raising us to a purer life and higher morality—bring ourselves one step nearer to salvation. It is true this is flatly contradicted by the previous law, which attributes the loss of salvation to *neglect*; but such trifles as flat contradictions do not much affect those who attempt to "read science into religion," and they easily escape detection if wrapped up in long scientific words and lubricated by an unctuous theology.

Death is the subject of the next chapter, and the argument is that, as death may be considered in the last resort to be the ceasing to be in correspondence with the environment, it is the inevitable fate of all who, not having been led by Divine grace to adopt the Calvinistic creed, are not in harmony with God; while, on the other hand, eternal life is the necessary attribute of all who have thus been brought into harmony with an eternal and unchanging environment.

It is wonderful how high-sounding theories are apt to collapse when touched by the Ithuriel spear of plain English. This of death being the ceasing to correspond with the environment simply amounts to this: that if we had not died we should be still alive—a truism which does not advance us much towards a solution of the great problem of a future life. To the ordinary apprehension of ordinary men the question of a future life means this: shall we, after death, retain the consciousness, or personal identity, which in this life distinguishes each individual from the surrounding universe. The practical test most would try it by is—shall we be able to meet and recognise those whom we have loved and lost?

The only elements reason is able to supply towards this momentous question are that, as far as our experience and knowledge extend, this life of conscious personal identity is indissolubly connected with a material organ—the brain. It did not exist before we were born; it only came gradually into existence as the infant brain grew and received impressions; it is suspended when the action of the brain is suspended, as in dreamless sleep and suspended animation; it is strangely distorted or duplicated in abnormal conditions of the brain, as in trance or hypnotism. What will become of it when the brain is dissolved into its elements? No voice comes from beyond the grave to tell us. It is the mystery of mysteries.

Behind the veil, behind the veil!

It is simply childish to tell us that the unknown can be solved by any analogy, more or less fanciful and far-fetched, to the natural laws which bind together phenomena which we really do know. Because matter cannot be created or destroyed, but only transformed, what does this tell us as to whether personal identity will be continued after death, or annihilated, or absorbed in the great ocean of an all-pervading spirit?

The next chapter is on mortification. This hardly takes the form of scientific argument, but is substantially a sermon on the text of "If thine eye offend thee, pluck it out." As far as any argument goes, the inference, as stated by Professor Drummond himself, seems to be that the best course would be for a man, directly he felt the vivifying influence of Divine grace, to commit suicide, and thus escape from the old environment of the natural man into the safe refuge of eternal life. If, in condescension to human weakness, this extreme remedy is not adopted, the next best course is "to die as much as he can," and withdraw from all the duties, affections, interests, and pleasures of natural life, into a rigid asceticism.

To become a Christian fakir is the ideal set before us for those who have not the courage to adopt the more complete remedy of suicide.

The chapter on eternal life is a continuation of the same argument as that on death.

If Herbert Spencer, in a philosophical discussion on life and death, tells us that with an eternal correspondence between an organism and its environment the organism would live for ever, he simply tells us, in abstract terms, that if there were no cause for death we should continue to live. This, Professor Drummond calls "one of the most startling achievements of recent science, and a contribution of immense moment to the religious mind."

No one would probably be more surprised than Mr. Herbert Spencer to find that this generalisation of his had been accepted as a positive scientific proof of the "Shorter Catechism." The Professor is much too apt to forget the sage aphorism which applies to philosophical and theological speculations, as well as to more sublunary matters: "First catch your hare." First *prove* the reality of your spiritual life, and it will be time to consider the different scientific sauces with which it may be dressed up.

In what possible way does Mr. Herbert Spencer's generalisation affect the question whether the Bible is inspired; whether it is a true revelation of things otherwise unknowable; and if it be, whether the Calvinistic creed is the true interpretation of it? All these are questions which require to be established by solid proof, before we can even enter on the discussion of whether anything can be found in scientific laws or philosophical definitions, which may be thought to afford a more or less fanciful analogy to its peculiar dogmas.

After eternal life comes environment, and here, perhaps, the contrast between the scientific lecturer and the popular preacher comes out more sharply than in any other chapter. The first half is taken up by enumerating instances of the dependence of organisms on their environment. He shows how the colour of animals is modified by their surroundings; how the polar bear is white, the tiger striped, the flounder of sandy hue; how, without air, there could be no mammals, without water no fish, without environment no life. And then he jumps at once to this astounding corollary, that these facts are a mere scientific re-statement of the saying of Christ: "Without me ye can do nothing;" and the rest of the chapter is very much in the tone of an ordinary sermon on the text of the "lilies of the field," or, "take no thought for the morrow," inculcating absolute dependence on the spiritual environment, "which is God."

Conformity to type.

The scientific portion of this chapter is based on the fact that, within a limited range of time types breed true, and species of animal life are distinguished from one another by differences which remain constant and admit of classification. The theological inference drawn is that, "As the bird-life builds up a bird the image of itself, so the Christ-life builds up a Christ the image of Himself, in the inward nature of man."

The practical conclusion is that this establishes the doctrine of predestination. "Whom He did foreknow, He also did predestinate to be conformed to the image of His Son."

He adds: "One must confess that the originality of the entire New Testament conception is most startling."

No wonder, for to any ordinary mind it must appear startling to be told that predestination is a certain fact because dogs are not bred from birds' eggs.

To establish even the faintest analogy to the Christ-life which is assumed, it would be requisite to prove that higher types have invariably been evolved from lower ones, by some miraculous influence transforming at once a certain number of favoured individuals. Directly the contrary is known to be the case.

Types have arisen, flourished, in some cases decayed and died out, in others been transformed, not by any sudden process, but by the slow accumulation over long periods of time, of individual peculiarities, accumulated and fixed by the action of heredity and environment. Bird-life was not always bird-life ; it began as reptilian life, and the Archœopteryx is more of a lizard than of a bird.

If "conformity to type" really taught anything, it would tell rather in favour of death than of life, for it is certain that many highly organised types of life have died out and disappeared during past geological ages, and science, in the case of the moon, which being a smaller body than the earth has gone through its course of evolution quicker, points rather to ultimate death than to the passage into a higher stage of existence, of all suns, planets, and their inhabitants. But it would be as unscientific to draw conclusions from this, or from the law by which all energy tends to run down into one uniform ocean of rest, as temperatures become equalised, in favour of death as the law of the Unknown, as it is for Professor Drummond to draw from the same premises the conclusion of a Christ-life. It is either altogether unknown, or known only by revelation, and the first condition of the problem is to prove the revelation.

Parasitism and semi-parasitism.

These chapters give, in much detail, instances of the

natural law by which organisms who take life too easily and lean on others for support, degenerate and fall low in the scale of existence.

Thus the hermit crab, who is too lazy to make his own shell, and borrows the cast-off shell of some mollusk, loses the shell-secreting faculty, and falls behind the more laborious common crab. This is called semi-parasitism, while parasitism proper extends to the cases where the animal lives in another living animal, and degenerates into a mere sac, absorbing nourishment and laying eggs.

The conclusions drawn from this collection of interesting facts are certainly most extraordinary. " Roman Catholicism is an organisation specially designed to induce the parasitic habit in the souls of men. It offers the masses a molluscous shell." Even more startling it is to be told that " one of the things in the religious world which tends most strongly to induce the parasitic habit is *going to church*." The italics are not mine but the Professor's. And again : " In those churches, especially when all parts of the worship are subordinated to the sermon, this species of parasitism is peculiarly encouraged."

Nay, more, the better the preacher the greater is the danger, and if " Providence had not mercifully delivered the Church from too many great men in its pulpits," the consequences would have been most disastrous to a large circle of Christian people. Church-going Christians may perhaps find some consolation in the obvious fact, that if parasitism be such a deadly danger its extremest form would be found in the very spiritual life which Professor Drummond is attempting to prove. A more complete analogy to the parasitic sac cannot be found than that of the man who, fastening on to the Calvinistic creed, and arriving at the conviction that he is one of the elect, proceeds, as the Professor advises, "to die as much as he can," and abstracts himself from all the interests and duties of his natural environment.

The chapters are chiefly interesting as showing the length to which a learned and sincere man, who starts from the predetermination to believe a particular creed, can go on inventing arguments in its support, which, if they were worth anything, would really be most conclusive against it.

Classification.—The argument of this last chapter is not very apparent. No doubt all the facts of the inorganic and organic worlds, and those relating to natural man, admit of being arranged and classified. Religions also may be classified so far as they relate to known facts. Thus Mahometanism and Christianity may be classified as two of the world's religions, for there is no doubt of the fact that there are many millions both of Mahometans and of Christians. Or, again, religions may be classified as monotheistic or polytheistic, for, as a matter of fact, both have existed. But this tells us nothing of their intrinsic or relative truth.

So, if we assume the existence of Professor Drummond's spiritual world, those who belong to it, or even without assuming its existence, those who believe in it, may fairly be classed as a distinct sect from the rest of mankind. But this no more proves its reality than the classification of negroes as fetish-worshippers proves the truth of fetish worship. As usual, he has to fall back on texts, and quotes from St. John and St. Paul, sayings which seem to establish the reality of a wide distinction between carnal and spiritual life.

It might fairly be asked how we can be certain that many of these sayings are not merely the highly-coloured metaphorical expressions in which the Eastern mind invariably clothes its ideas, and whether they ought to be taken in the strict and literal sense, which the words present to the more practical and scientific European intellect.

But apart from this question, how does the fact that natural phenomena admit of classification, advance in the slightest degree the proposition that, in addition to the known inorganic and organic kingdoms, there must be a third unknown kingdom, which may be best designated as the "Kingdom of God?" There may or may not be such a kingdom, but assuredly, apart from revelation, we can no more prove or disprove from natural laws, that we shall live after death, than we can that we have lived before birth.

It would be easy, taking each chapter in detail, to show the fallacies involved in many of the analogies, and the extent to which scientific facts have been disturbed by the preconceived determination to make them square with the theory of a "Spiritual Life." For instance, when in order to prove the doctrine of eternal life, we are told, "that as we ascend in the scale of life we also rise in the scale of longevity," forgetting that, in this case, the parrot and the tortoise would take precedence of man as heirs of immortality.

Or again, when to prove original sin and redemption, we are told that there is in human nature a principle constantly dragging it down to a lower level, which can only be counteracted by the Christ-life; forgetting that long before Christ appeared, humanity had risen, intellectually, from the fabrication of stone hatchets to the perfection of tools and technical skill shown in the pyramids; and morally, from the cannibal feasts of the cavern of Chaleux to the ethics of a Socrates and a Plato.

But objections of detail are irrelevant, when it is so obvious that the whole edifice of Professor Drummond's superstructure rests on the *assumption* that the spiritual life of his definition is a proved and undoubted fact.

This again rests on the *assumption* that certain texts, quoted almost entirely from the writings of two of the many writers whose works constitute the Bible, St. Paul and St. John, are inspired revelations of the word of God, and therefore absolutely and certainly true.

Take this away, and nothing remains of the peculiar "Spiritual World" and "Christ-life," which are the axioms upon which he builds up every one of his supposed analogies to natural laws. For we can hardly call proof the assertion that these axioms are self-evident to what he admits to be an almost infinitesimally small portion of the whole world, and even of the Christian world. If this were proof it would apply equally to every religion and every superstition, or sect of religion, that has ever existed in the world.

And in the same manner the analogies would apply as well, or in many cases better, to other totally different forms of religious belief.

This has been already shown generally of his main proposition, and it can be shown in detail of each one of the natural laws which form the subject of the separate chapters.

For instance, those of degeneration and parasitism fit in far better with what may be called the Catholic

Christianity of the great majority, which places works above faith, and seeks to rise to a higher level by strenuous and persistent effort, than with a theory which makes salvation depend on a sudden miraculous act of Divine grace, fixed by predestination, or "blowing where it listeth."

Or, if a learned Brahmin or Buddhist read the chapter on mortification, he would exclaim: "Why, here is my faith, and the essence of my religion." Why does the holy fakir sit naked in the rain and wind, with his hands clasped till the nails grow through the flesh, or upraised till the muscles become rigid, if it be not to "die as much as he can," detach himself from the evil environment of the natural world, and so anticipate the time when his little rill of illusive individual existence may be absorbed in the mighty ocean of the universal Spirit?

And so of each of the chapters. Better analogies could readily be found for each of them in other creeds; better, because they would not be mutually contradictory, as these are in assigning in one place persistent effort, and in another, asceticism and passive acquiescence in predestined grace, as the conditions of attaining spiritual life.

The truth is, as we have already said, that Professor Drummond, like so many other theological writers, begins at the wrong end.

There is absolutely no foundation for his superstructure, except in the assured belief:

First, in revelation as taught mankind by an inspired book;

Secondly, in the particular interpretation given to it by the Calvinistic creed.

Let him begin at the beginning, and lay the foundation stone, solidly and securely, and it will be time to examine whether the edifice he has built upon it is likely to stand, or is destined to be one of the many enthusiastic speculations, which, in his own words, speaking of his own creed, "rise into prominence from time to time, become the watchwords of insignificant parties, and die down ultimately for want of lives to live them."

THE END.

MODERN SCIENCE AND MODERN THOUGHT.

By S. LAING.

THE TIMES.

"Mr. Laing is a man of active mind, and he has had a busy and rather multifarious life. He is on good terms with his work, his fellow-workers, and his fellow-thinkers. He treads a beaten path. If he does not pretend to originality it is because science is not original except to those who devote themselves to some one field of investigation. He reports what is known, or believed or not believed, by the majority of scientific men. The character of the work is foreshadowed in its divisions and titles. Two hundred pages are given to 'Science,' followed by about one hundred professedly given to 'Thought.' The thought, however, is scientific, and it is science that dominates from the first page to the last. In the first part Mr. Laing exhibits with much power and effect the immense discoveries of Science, and its numerous victories over old opinions whenever they have had the rashness to challenge conclusions with it. These discoveries are not so familiar to the world at large but that any ordinary reader may learn much from a writer combining matter and style, and conveying solid information in simple yet striking language. In a comparatively small compass are here displayed the results of recent inquiries into the composition and constitution of the earth and of the universe, into the nature and laws of matter, into the development of organised and animated existence, into the history of man, into the myths of all races and the faiths of all people; into force, motion, electricity, light, and heat. As one turns over the glowing pages one is tempted to lament that a man so qualified to instruct and to illustrate should have been almost exclusively occupied in absorbing official and practical duties."

THE PALL MALL GAZETTE.

"Apart from the uselessness and undue multiplication of all such books, Mr. Laing's brief statement of an agnostic creed is good enough and sensible enough in its own way. It is the expression of a sensible, well-read, compromise-loving Briton's final conclusions upon religious matters. The first part is a rapid and clearly written résumé of all that Modern Science and Modern Criticism have done to sap the foundation of current theologies and the current dogmas. This résumé is admirably done. Mr. Laing manages to condense into a few short chapters an amount of salient information on matters astronomical, geological, archæological, and historical; and withal he condenses it cleverly. . . . The evidence of geology against the Mosaic cosmogony; the evidence of biology, and especially of evolutionism, against the story of creation; the evidence of the palæolithic flints and the reindeer age cave-men against the naïve history of Adam and Eve; the evidence of human development against the entire Biblical conception of man's importance in the scheme of nature—all marshalled with considerable skill, and enforced by excellent and typical examples. The anxious but unlearned inquirer who really wishes to know how much recent researches have effected towards undermining the groundwork of the existing creeds, cannot do better than turn to Mr. Laing's pleasantly written pages."

THE SCOTSMAN.

"In his first part Mr. Laing has presented the chief results of modern scientific investigation with singular terseness and lucidity, and what is more, he has contrived to indicate by simple and impressive illustrations the methods by which these results have been attained. And this he has done in a style so simple and elementary, yet so sufficient for his purpose, that any fairly intelligent reader who has never been able to give attention to scientific subjects may fully grasp the sum of the knowledge he seeks to impart. . . . The chapters of the second part are of genuine interest as showing the conclusions of a practical man who belongs neither to the philosophers nor the theologians, but has intelligently studied the researches and reasonings of both, and has formed his judgment between them on the principles of common sense, and by means of the ordinary rules by which men weigh evidence in ordinary life."

THE WESTMINSTER REVIEW.

"From the first page to the last the book is charmingly written, with temperance and wisdom that will win a hearing for the author from many who may not share his views."

JOURNAL OF SCIENCE.

The successive chapters of the first part of the work discuss Space, Time, Matter, Life, the Antiquity of Man, and Man's place in Nature. As a whole the picture given of the Universe and of its development is clear, and on a level with the present state of human knowledge."

LANCET.

"This work is something more than a compilation. It is true that the author has contributed nothing new to Science, but he has filtered a large amount of information, and the outcome is that he has admirably selected and arranged his material, and added to it the impressions it has made upon a highly cultivated intellect. The author's style is clear and terse, and the work as an exposition of Modern Science is decidedly interesting."

ACADEMY.

'It would be unjust to conclude without acknowledging that Mr. Laing is always candid and generally accurate, that he writes a clear and vigorous style, and that he has brought together a number of facts and arguments which will be studied with interest both by those who go farther than he does, and by those who do not go so far."

NATIONAL REFORMER.

"This handsome and most interesting volume purports to give a clear and concise view of the principal results of modern science and of the revolution which they have effected in Modern Thought. It is divided into two parts: the first showing the results of very close and careful research, presenting the teachings of modern science on Space, Time, Matter, Life, the Antiquity of Man, and Man's place in Nature; and the second part dealing with Modern Thought, Miracles, Christianity without Miracles, and Practical Life."

BIRMINGHAM DAILY POST.

"Such a work as this of Mr. Laing's may be accepted with thankfulness. In the first part he presents concisely the principal results of Modern Science and of the revolutions which they have effected in Modern Thought. This part of the work the author alludes to with so much modesty that those who take aim at his own valuation will scarcely realise what is involved in presenting a clear, concise view of the results of Modern Science. It means that many weighty and thoughtful books have been studied and mastered, their contents assimilated. It means a logical and orderly mind capable of arranging infinite details of a complex and disconnected character in a harmonious and orderly whole. This task has been carried out with judgment in the selection of facts to be presented and lucidity of presentation. The growth of the ideas of Space, Time, Matter—the attempts to gather the secret of the great mystery of life—the Antiquity of Man, and Man's place in Nature, these are traced and examined, the light which the researches of Lyell and Lubbock, Huxley, Proctor, and Darwin have thrown on them is displayed. Thus 'the common property of thinking minds' is set forth in eminently attractive prose."

SPECTATOR.

"By far the greater portion of the book consists of a summary of contemporary Science as bearing upon the great questions of Time, Space, Matter, Life, and Man. . . . His last chapter on 'Practical Life' is full of wisdom which is not wholly worldly."

SCOTTISH REVIEW.

"Among readers of a more robust and sceptical tendency this volume of Mr. Laing's seems likely to take a place somewhat analogous to that which Mr. Drummond's 'Natural Law in the Spiritual World' has acquired among their orthodox brethren. Both books deal largely with Science, and both mix up Science with Theology and Religion. It is clear, logical, straightforward, and remarkably outspoken."

POPULAR SCIENCE MONTHLY.
NEW YORK.

"Modern Science and Modern Thought" is a remarkable and vigorous article from a new English work, by S. Laing. The liberal tendencies of modern opinion following the revolution of scientific ideas are presented in a very effective manner."

CHARLES DICKENS AND EVANS, CRYSTAL PALACE PRESS.

11, Henrietta Street, Covent Garden, W.C.

April, 1889.

A Catalogue of Books

PUBLISHED BY

CHAPMAN & HALL,

LIMITED.

FOR

Drawing Examples, Diagrams, Models, Instruments, etc.,

ISSUED UNDER THE AUTHORITY OF

THE SCIENCE AND ART DEPARTMENT,
SOUTH KENSINGTON,

FOR THE USE OF SCHOOLS AND ART AND SCIENCE CLASSES,

See separate Illustrated Catalogue.

NEW BOOKS FOR APRIL.

TEN YEARS' WILD SPORTS IN FOREIGN LANDS;
Or, Travels in the Eighties. By H. W. Seton-Karr, F.R.G.S., etc. Demy 8vo.

MADAME DE STAËL: Her Friends, and Her Influence
in Politics and Literature. By Lady Blennerhassett. With a Portrait. 3 vols. Demy 8vo, 36s.

HISTORY OF THE PEOPLE OF ISRAEL. From the Reign of David up to the Capture of Samaria. By Ernest Renan. Second Division. Demy 8vo, 14s.

FROM PEKIN TO CALAIS BY LAND. By H. de Windt. With numerous Illustrations by C. E. Fripp from Sketches by the Author. Demy 8vo, 20s.

THE HISTORY OF ANCIENT CIVILISATION.
Handbook based upon M. Gustave Ducoudray's "Histoire Sommaire de la Civilisation." Edited by Rev. J. Verschoyle, M.A. With Illustrations. Large crown 8vo, 6s.

HALF A CENTURY OF MUSIC IN ENGLAND.
1837—1887. By F. Hueffer, Author of "Richard Wagner and the Music of the Future." Demy 8vo.

THE MARRIAGES OF THE BOURBONS. By Capt. the Hon. D. A. Bingham. 2 vols. Demy 8vo.

GALILEO AND HIS JUDGES. By F. R. Wegg-Prosser. Demy 8vo, 5s.

THE LIFE OF THE RIGHT HON. W. E. FORSTER.
By T. Wemyss Reid. Fifth Edition. In 1 vol. Demy 8vo, 10s. 6d.

GIBRALTAR. By Henry M. Field. With numerous Illustrations. Demy 8vo.

THE SALMON AND ITS HABITS. By Major Traherne. Crown 8vo.

A SUBURB OF YEDO. By T. A. P. With Illustrations. Crown 8vo.

BOOKS

PUBLISHED BY

CHAPMAN & HALL, LIMITED.

ABLETT (T. R.)—
   WRITTEN DESIGN. Oblong, sewed, 6d.

ABOUT (EDMOND)—
   HANDBOOK OF SOCIAL ECONOMY; OR, THE WORKER'S A B C. From the French. With a Biographical and Critical Introduction by W. Fraser Rae. Second Edition, revised. Crown 8vo, 4s.

   AFRICAN FARM, STORY OF AN. By Olive Schreiner (Ralph Iron). New Edition. Crown 8vo, 1s.; in cloth, 2s.

ANDERSON (ANDREW A.)—
   TWENTY-FIVE YEARS IN A WAGGON IN THE GOLD REGIONS OF AFRICA. With Illustrations and Map. Second Edition. Demy 8vo, 12s.

   AGRICULTURAL SCIENCE (LECTURES ON), AND OTHER PROCEEDINGS OF THE INSTITUTE OF AGRICULTURE, SOUTH KENSINGTON, 1883-4. Crown 8vo, sewed, 2s.

AVELING (EDWARD), D.Sc., Fellow of University College, London—
   MECHANICS AND EXPERIMENTAL SCIENCE. As required for the Matriculation Examination of the University of London.
      MECHANICS. With numerous Woodcuts. Crown 8vo, 6s.
         Key to Problems in ditto, crown 8vo, 3s. 6d.
      CHEMISTRY. With numerous Woodcuts. Crown 8vo, 6s.
      MAGNETISM AND ELECTRICITY. With Numerous Woodcuts. Crown 8vo, 6s.
      LIGHT AND HEAT. With Numerous Woodcuts. Crown 8vo, 6s.
         Keys to above volumes in the Press.

BADEN-POWELL (GEORGE)—
   STATE AID AND STATE INTERFERENCE. Illustrated by Results in Commerce and Industry. Crown 8vo, 9s.

BAILEY (JOHN BURN)—
   MODERN METHUSELAHS; or, Short Biographical Sketches of a few advanced Nonagenarians or actual Centenarians who were distinguished in Art, Science, or Philanthropy. Also brief notices of some individuals remarkable chiefly for their longevity. With an Introductory Chapter on "Long-Lasting." Demy 8vo, 10s. 6d.

BARKER (G. F. RUSSELL) and DAUGLISH (M. G.), of Lincoln's Inn, Barristers-at-Law—
   HISTORICAL AND POLITICAL HANDBOOK. Second Edition. Crown 8vo, 2s. 6d.

BARTLEY (G. C. T.)—
A HANDY BOOK FOR GUARDIANS OF THE POOR.
Crown 8vo, cloth, 3s.
BAYARD: HISTORY OF THE GOOD CHEVALIER,
SANS PEUR ET SANS REPROCHE. Compiled by the LOYAL SERVITEUR. With over 200 Illustrations. Royal 8vo, 21s.

BEATTY-KINGSTON (W.)—
A WANDERER'S NOTES. 2 vols. Demy 8vo, 24s.
MONARCHS I HAVE MET. 2 vols. Demy 8vo, 24s.
MUSIC AND MANNERS: Personal Reminiscences and
Sketches of Character. 2 vols. Demy 8vo, 30s.

BELL (JAMES, Ph.D., &c.), Principal of the Somerset House Laboratory—
THE CHEMISTRY OF FOODS. With Microscopic
Illustrations.
PART I. TEA, COFFEE, COCOA, SUGAR, ETC. Large crown 8vo, 2s. 6d.
PART II. MILK, BUTTER, CHEESE, CEREALS, PREPARED STARCHES, ETC. Large crown 8vo, 3s.

BENSON (W.)—
UNIVERSAL PHONOGRAPHY. To classify sounds of
Human Speech, and to denote them by one set of Symbols for easy Writing and Printing. 8vo, sewed, 1s.
MANUAL OF THE SCIENCE OF COLOUR. Coloured
Frontispiece and Illustrations. 12mo, cloth, 2s. 6d.
PRINCIPLES OF THE SCIENCE OF COLOUR. Small
4to, cloth, 15s.

GHAM (CAPT. THE HON. D.)—
A SELECTION FROM THE LETTERS AND
DESPATCHES OF THE FIRST NAPOLEON. With Explanatory Notes. 3 vols. Demy 8vo, £2 2s.
THE BASTILLE. With Illustrations. 2 vols. Demy 8vo,
30s.
THE MARRIAGES OF THE BOURBONS. 2 vols.
Demy 8vo. [*In the Press.*

BIRDWOOD (SIR GEORGE C. M.), C.S.I.—
THE INDUSTRIAL ARTS OF INDIA. With Map and
174 Illustrations. New Edition. Demy 8vo, 14s.

BLACKIE (JOHN STUART), F.R.S.E.—
THE SCOTTISH HIGHLANDERS AND THE LAND
LAWS. Demy 8vo, 9s.
ALTAVONA: FACT AND FICTION FROM MY LIFE
IN THE HIGHLANDS. Third Edition. Crown 8vo, 6s.

BLATHERWICK (CHARLES)—
PERSONAL RECOLLECTIONS OF PETER STONNOR,
Esq. With Illustrations by JAMES GUTHRIE and A. S. BOYD. Large crown 8vo, 6s.

BLOOMFIELD'S (BENJAMIN LORD), MEMOIR OF—
MISSION TO THE COURT OF BERNADOTTE. Edited by GEORGIANA, BARONESS BLOOMFIELD, Author of "Reminiscences of Court and Diplomatic Life." With Portraits. 2 vols. Demy 8vo, 28s.

BLENNERHASSETT (LADY).—
MADAME DE STAËL: Her Friends, and Her Influence in Politics and Literature. With a Portrait. 3 vols. Demy 8vo, 36s.

BONVALOT (GABRIEL)—
THROUGH THE HEART OF ASIA OVER THE PAMIR TO INDIA. Translated from the French by C. B. PITMAN. With 250 Illustrations by ALBERT PÉPIN. Royal 8vo, 32s.

BOULGER (DEMETRIUS C.)—
GENERAL GORDON'S LETTERS FROM THE CRIMEA, THE DANUBE, AND ARMENIA. 2nd Edition. Crown 8vo, 5s.

BOWERS (G.)—
HUNTING IN HARD TIMES. With 61 coloured Illustrations. Oblong 4to, 12s.

BRACKENBURY (COL. C. B.)—
FREDERICK THE GREAT. With Maps and Portrait. Large crown 8vo, 4s.

BRADLEY (THOMAS), of the Royal Military Academy, Woolwich—
ELEMENTS OF GEOMETRICAL DRAWING. In Two Parts, with Sixty Plates. Oblong folio, half bound, each Part 16s.

MRS. BRAY'S NOVELS AND ROMANCES.
New and Revised Editions, with Frontispieces. 3s. 6d. each.

THE WHITE HOODS; a Romance of Flanders.
DE FOIX; a Romance of Bearn.
THE TALBA; or, The Moor of Portugal.
THE PROTESTANT; a Tale of the Times of Queen Mary.

NOVELS FOUNDED ON TRADITIONS OF DEVON AND CORNWALL.

FITZ OF FITZFORD; a Tale of Destiny.
HENRY DE POMEROY; or, the Eve of St. John.
TRELAWNY OF TRELAWNE; or, a Romance of the West.
WARLEIGH; or, The Fatal Oak.
COURTENAY OF WALREDDON; a Romance of the West.
HARTLAND FOREST AND ROSE-TEAGUE.

MISCELLANEOUS TALES.
A FATHER'S CURSE AND A DAUGHTER'S SACRIFICE.
TRIALS OF THE HEART.

BRITISH ARMY, THE. By the Author of "Greater Britain," "The Present Position of European Politics," etc. Demy 8vo, 12s.

BROADLEY (A. M.)—
HOW WE DEFENDED ARABI AND HIS FRIENDS. A Story of Egypt and the Egyptians. Illustrated by FREDERICK VILLIERS. Demy 8vo, 12s.

BROMLEY-DAVENPORT (the late W.), M.P.—
SPORT: Fox Hunting, Salmon Fishing, Covert Shooting, Deer Stalking. With numerous Illustrations by General CREALOCK, C.B. New Cheap Edition. Post 8vo, 3s. 6d.

——— Small 4to, 21s.

BUCKLAND (FRANK)—
LOG-BOOK OF A FISHERMAN AND ZOOLOGIST. With numerous Illustrations. Fifth Thousand. Crown 8vo, 5s.

BROWN (J. MORAY)—
  POWDER, SPEAR, AND SPUR: A Sporting Medley.
  With Illustrations by G. D. Giles and Edgar Giberne from Sketches by the Author. Crown 8vo, 10s. 6d.

BURCHETT (R.)—
  DEFINITIONS OF GEOMETRY. New Edition. 24mo, cloth, 5d.

  LINEAR PERSPECTIVE, for the Use of Schools of Art.
  New Edition. With Illustrations. Post 8vo, cloth, 7s.

  PRACTICAL GEOMETRY: The Course of Construction of Plane Geometrical Figures. With 137 Diagrams. Eighteenth Edition. Post 8vo, cloth, 5s.

BURGESS (EDWARD)—
  ENGLISH AND AMERICAN YACHTS. Illustrating and Describing the most famous Yachts now sailing in English and American Waters. With a treatise upon Yachts and Yachting. Illustrated with 50 Beautiful Photogravure Engravings. Oblong folio, 42s.

BUTLER (A. J.)—
  COURT LIFE IN EGYPT. Second Edition. Illustrated.
  Large crown 8vo, 12s.

CARLYLE (THOMAS), WORKS BY.—See pages 29 and 30.
  THE CARLYLE BIRTHDAY BOOK. Compiled, with the permission of Mr. Thomas Carlyle, by C. N. Williamson. Second Edition. Small fcap. 8vo, 3s.

CHALDÆAN AND ASSYRIAN ART—
  A HISTORY OF ART IN CHALDÆA AND ASSYRIA.
  By Georges Perrot and Charles Chipiez. Translated by Walter Armstrong, B.A. Oxon. With 452 Illustrations. 2 vols. Imperial 8vo, 42s.

CHARNAY (DÉSIRÉ)—
  THE ANCIENT CITIES OF THE NEW WORLD.
  Being Travels and Explorations in Mexico and Central America, 1857—1882. Translated from the French by J. Gonino and Helen S. Conant. With upwards of 200 Illustrations. Super Royal 8vo, 31s. 6d.

CHURCH (PROFESSOR A. H.), M.A. Oxon.—
  FOOD GRAINS OF INDIA. With numerous Woodcuts.
  Small 4to, 6s.

  ENGLISH PORCELAIN. A Handbook to the China made in England during the Eighteenth Century, as illustrated by Specimens chiefly in the National Collection. With numerous Woodcuts. Large crown 8vo, 3s.

  ENGLISH EARTHENWARE. A Handbook to the Wares made in England during the 17th and 18th Centuries, as illustrated by Specimens in the National Collections. With numerous Woodcuts. Large crown 8vo, 3s.

  PLAIN WORDS ABOUT WATER. Illustrated. Crown 8vo, sewed, 6d.

CHURCH (*PROFESSOR A. H.*), *M.A. Oxon.* (*Continued*)—
  FOOD: Some Account of its Sources, Constituents, and Uses. Sixth Thousand. Large crown 8vo, cloth, 3s.
  PRECIOUS STONES: considered in their Scientific and Artistic Relations. With a Catalogue of the Townsend Collection of Gems in the South Kensington Museum. With a Coloured Plate and Woodcuts. Large crown 8vo, 2s. 6d.

CLINTON (*R. H.*)—
  A COMPENDIUM OF ENGLISH HISTORY, from the Earliest Times to A.D. 1872. With Copious Quotations on the Leading Events and the Constitutional History, together with Appendices. Post 8vo, 7s. 6d.

COBDEN, RICHARD, LIFE OF. By the RIGHT HON. JOHN MORLEY, M.P. With Portrait. New Edition. Crown 8vo, 7s. 6d.
  Popular Edition, with Portrait, 4to, sewed, 1s.; cloth, 2s.

COOKERY—
  THE PYTCHLEY BOOK OF REFINED COOKERY AND BILLS OF FARE. By MAJOR L——. Second Edition. Large crown 8vo, 8s.
  BREAKFASTS, LUNCHEONS, AND BALL SUPPERS. By MAJOR L——. Crown 8vo, 4s.
  OFFICIAL HANDBOOK OF THE NATIONAL TRAINING SCHOOL FOR COOKERY. Containing Lessons on Cookery; forming the Course of Instruction in the School. Compiled by "R. O. C." Eighteenth Thousand. Large crown 8vo, 6s.
  BREAKFAST AND SAVOURY DISHES. By "R. O. C." Seventh Thousand. Crown 8vo, 1s.
  HOW TO COOK FISH. Compiled by "R. O. C." Crown 8vo, sewed, 3d.
  SICK-ROOM COOKERY. Compiled by "R. O. C." Crown 8vo, sewed, 6d.
  THE ROYAL CONFECTIONER: English and Foreign. A Practical Treatise. By C. E. FRANCATELLI. With numerous Illustrations. Fifth Thousand. Crown 8vo, 5s.
  THE KINGSWOOD COOKERY BOOK. By H. F. WICKEN. Crown 8vo, 2s.

COOPER-KING (*LT.-COL.*)—
  GEORGE WASHINGTON. Large crown 8vo. With Portrait and Maps. [*In the Press.*

COURTNEY (*W. L.*), *M.A., LL.D., of New College, Oxford*—
  STUDIES NEW AND OLD. Crown 8vo, 6s.
  CONSTRUCTIVE ETHICS: A Review of Modern Philosophy and its Three Stages of Interpretation, Criticism, and Reconstruction. Demy 8vo, 12s.

CRAIK (*GEORGE LILLIE*)—
  ENGLISH OF SHAKESPEARE. Illustrated in a Philological Commentary on his "Julius Cæsar." Seventh Edition. Post 8vo, cloth, 5s.
  OUTLINES OF THE HISTORY OF THE ENGLISH LANGUAGE. Tenth Edition. Post 8vo, cloth, 2s. 6d.

CRAWFURD (OSWALD)—
 BEYOND THE SEAS; being the surprising Adventures and ingenious Opinions of Ralph, Lord St. Keyne, told by his kinsman, Humphrey St. Keyne. Second Edition. Crown 8vo, 3s. 6d.

CRIPPS (WILFRED JOSEPH), M.A., F.S.A.—
 COLLEGE AND CORPORATION PLATE. A Handbook for the Reproduction of Silver Plate. [*In the South Kensington Museum, from celebrated English collections.*] With numerous Illustrations. Large crown 8vo, cloth. 2s. 6d.

DAIRY FARMING—
 DAIRY FARMING. To which is added a Description of the Chief Continental Systems. With numerous Illustrations. By JAMES LONG. Crown 8vo, 9s.
 DAIRY FARMING, MANAGEMENT OF COWS, &c. By ARTHUR ROLAND. Edited by WILLIAM ABLETT. Crown 8vo, 5s.

DALY (J. B.), LL.D.—
 IRELAND IN THE DAYS OF DEAN SWIFT. Crown 8vo, 5s.

DAUBOURG (E.)—
 INTERIOR ARCHITECTURE. Doors, Vestibules, Staircases, Anterooms, Drawing, Dining, and Bed Rooms, Libraries, Bank and Newspaper Offices, Shop Fronts and Interiors. Half-imperial, cloth, £2 12s. 6d.

DAVIDSON (ELLIS A.)—
 PRETTY ARTS FOR THE EMPLOYMENT OF LEISURE HOURS. A Book for Ladies. With Illustrations. Demy 8vo, 6s.

DAVITT (MICHAEL)—
 LEAVES FROM A PRISON DIARY; or, Lectures to a Solitary Audience. Crown 8vo, 6s.
 Cheap Edition. Ninth Thousand. **Crown 8vo, sewed, 1s. 6d.**

DAY (WILLIAM)—
 THE RACEHORSE IN TRAINING, with Hints on Racing and Racing Reform, to which is added a Chapter on Shoeing. Sixth Edition. Demy 8vo, 9s.

DAS (DEVENDRA N.)—
 SKETCHES OF HINDOO LIFE. Crown 8vo, 5s.

DE AINSLIE (GENERAL)—
 A HISTORY OF THE ROYAL REGIMENT OF DRAGOONS. From its Formation in 1661 to the Present Day. With Illustrations. Demy 8vo, 21s.

DE CHAMPEAUX (ALFRED)—
 TAPESTRY. With numerous Woodcuts. Cloth, 2s. 6d.

DE FALLOUX (THE COUNT)—
 MEMOIRS OF A ROYALIST. Edited by C. B. PITMAN. 2 vols. With Portraits. Demy 8vo, 32s.

D'HAUSSONVILLE (VICOMTE)—
 SALON OF MADAME NECKER. Translated by H. M. TROLLOPE. 2 vols. Crown 8vo, 18s.

DE KONINCK (L. L.) and DIETZ (E.)—
    PRACTICAL MANUAL OF CHEMICAL ASSAYING,
    as applied to the Manufacture of Iron. Edited, with notes, by ROBERT MALLET. Post 8vo, cloth, 6s.

DE LESSEPS (FERDINAND)—
    RECOLLECTIONS OF FORTY YEARS. Translated from the French by C. B. PITMAN. 2 vols. Demy 8vo, 24s.

    DE LISLE (MEMOIR OF LIEUTENANT RUDOLPH), R.N., of the Naval Brigade. By the Rev. H. N. OXENHAM, M.A. Third Edition. Crown 8vo, 7s. 6d.

DE MANDAT-GRANCEY (BARON E.)—
    PADDY AT HOME; OR, IRELAND AND THE IRISH AT THE PRESENT TIME, AS SEEN BY A FRENCHMAN. Translated from the French. Fourth Edition. Crown 8vo, 7s.

DE STAËL (MADAME)—
    MADAME DE STAËL: Her Friends, and Her Influence in Politics and Literature. By LADY BLENNERHASSETT. With a Portrait. 3 vols. Demy 8vo, 36s.

DE WINDT (H.)—
    FROM PEKIN TO CALAIS BY LAND. With numerous Illustrations by C. E. FRIPP from Sketches by the Author. Demy 8vo, 20s.

DICKENS (CHARLES), WORKS BY—See pages 31—37.
    THE LETTERS OF CHARLES DICKENS. Two vols. uniform with "The Charles Dickens Edition" of his Works. Crown 8vo, 8s.

    THE LIFE OF CHARLES DICKENS—*See "Forster."*

    THE CHARLES DICKENS BIRTHDAY BOOK. With Five Illustrations. In a handsome fcap. 4to volume, 12s.

    THE HUMOUR AND PATHOS OF CHARLES DICKENS. By CHARLES KENT. With Portrait. Crown 8vo, 6s.

DILKE (LADY)—
    ART IN THE MODERN STATE. With Facsimile. Demy 8vo, 9s.

DOUGLAS (JOHN)—
    SKETCH OF THE FIRST PRINCIPLES OF PHYSIOGRAPHY. With Maps and numerous Illustrations. Crown 8vo, 6s.

DOWN WITH ENGLAND. Translated from the French. With Maps. Crown 8vo, 1s.

DRAYSON (MAJOR-GENERAL A. W.), Late R.A., F.R.A.S.—
    THIRTY THOUSAND YEARS OF THE EARTH'S PAST HISTORY. Large Crown 8vo, 5s.

    EXPERIENCES OF A WOOLWICH PROFESSOR during Fifteen Years at the Royal Military Academy. Demy 8vo, 8s.

    THE CAUSE OF THE SUPPOSED PROPER MOTION OF THE FIXED STARS. Demy 8vo, cloth, 10s.

    PRACTICAL MILITARY SURVEYING AND SKETCHING. Fifth Edition. Post 8vo, cloth, 4s. 6d.

DREAMS BY A FRENCH FIRESIDE. Translated from the German by MARY O'CALLAGHAN. Illustrated by Fred Roe. Crown 8vo, 7s. 6d.

DUCOUDRAY (GUSTAVE)—
THE HISTORY OF ANCIENT CIVILISATION. A Handbook based upon M. Gustave Ducoudray's "Histoire Sommaire de la Civilisation." Edited by REV. J. VERSCHOYLE, M.A. With Illustrations. Large crown 8vo, 6s.

DUFFY (SIR CHARLES GAVAN), K.C.M.G.—
THE LEAGUE OF NORTH AND SOUTH. An Episode in Irish History, 1850-1854. Crown 8vo, 8s.

DYCE (WILLIAM), R.A.—
DRAWING-BOOK OF THE GOVERNMENT SCHOOL OF DESIGN; OR, ELEMENTARY OUTLINES OF ORNAMENT. Fifty selected Plates. Folio, sewed, 5s.; mounted, 18s.

ELEMENTARY OUTLINES OF ORNAMENT. Plates I. to XXII., containing 97 **Examples, adapted** for Practice of Standards **I. to IV.** Small folio, sewed, 2s. 6d.

SELECTION FROM DYCE'S DRAWING BOOK. 15 Plates, sewed, 1s. 6d.; mounted on cardboard, 6s. 6d.

TEXT TO ABOVE. Crown 8vo, sewed, **6d.**

EDWARDS (H. SUTHERLAND)—
FAMOUS FIRST REPRESENTATIONS. Crown 8vo, 6s.

EGYPTIAN ART—
A HISTORY OF ART IN ANCIENT EGYPT. By G. PERROT and C. CHIPIEZ. Translated by WALTER ARMSTRONG. With over 600 Illustrations. 2 vols. Imperial 8vo, £2 2s.

ELLIS (A. B., Major 1st West India Regiment)—
WEST AFRICAN STORIES. Crown 8vo.

THE TSHI-SPEAKING PEOPLES OF THE GOLD COAST OF WEST AFRICA: their Religion, Manners, Customs, Laws, Language, &c. With Map. Demy 8vo, 10s. 6d.

SOUTH **AFRICAN SKETCHES.** Crown 8vo, 6s.

WEST AFRICAN ISLANDS. Demy 8vo, 14s.

THE HISTORY OF THE WEST INDIA REGI- MENT. With Maps and Coloured Frontispiece and Title-page. Demy 8vo, 18s.

THE LAND OF FETISH. Demy 8vo, 12s.

ENGEL (CARL)—
MUSICAL INSTRUMENTS. With numerous Woodcuts. Large crown 8vo, cloth, 2s. 6d.

ESCOTT (T. H. S.)—
POLITICS AND LETTERS. Demy 8vo, 9s.

ENGLAND. ITS PEOPLE, POLITY, **AND** PURSUITS. New and Revised Edition. Sixth Thousand. 8vo, 8s.

EUROPEAN POLITICS, THE PRESENT POSITION OF. By **the** Author of "Greater Britain." Demy 8vo, 12s.

FANE (VIOLET)—
QUEEN OF THE FAIRIES (A Village Story), **and other** Poems. Crown 8vo, 6s.

ANTHONY BABINGTON: a Drama. Crown 8vo, 6s.

FARR (WILLIAM) and THRUPP (GEORGE A.)—
    COACH TRIMMING. With 60 Illustrations. Crown 8vo,
    2s. 6d.

FIELD (HENRY M.)—
    GIBRALTAR. With numerous Illustrations. Demy 8vo,
    7s. 6d.

FIFE-COOKSON (LIEUT.-COL. J. C.)—
    TIGER-SHOOTING IN THE DOON AND ULWAR,
    AND LIFE IN INDIA. With numerous Illustrations by E. HOBDAY, R.H.A.
    Large crown 8vo, 10s. 6d.

FITZGERALD (PERCY), F.S.A.—
    THE CHRONICLES OF BOW STREET POLICE
    OFFICE, with an Account of the Magistrates, "Runners," and Police; and a
    selection of the most interesting Cases. With numerous Illustrations. 2 vols.
    Demy 8vo, 21s.

FLEMING (GEORGE), F.R.C.S.—
    ANIMAL PLAGUES: THEIR HISTORY, NATURE,
    AND PREVENTION. 8vo, cloth, 15s.

    PRACTICAL HORSE-SHOEING. With 37 Illustrations.
    Fifth Edition, enlarged. 8vo, sewed, 2s.

    RABIES AND HYDROPHOBIA: THEIR HISTORY,
    NATURE, CAUSES, SYMPTOMS, AND PREVENTION. With 8 Illustrations. 8vo, cloth, 15s.

FLOYER (A. M.)—
    EVOLUTION OF ANCIENT HINDUISM. Crown 8vo,
    2s. 6d.

FORSTER (JOHN)—
    THE LIFE OF CHARLES DICKENS. Uniform with
    the Illustrated Library Edition of Dickens's Works. 2 vols. Demy 8vo, 20s.

    THE LIFE OF CHARLES DICKENS. Uniform with
    the Library Edition. Post 8vo, 10s. 6d.

    THE LIFE OF CHARLES DICKENS. Uniform with
    the "C. D." Edition. With Numerous Illustrations. 2 vols. 7s.

    THE LIFE OF CHARLES DICKENS. Uniform with
    the Household Edition. With Illustrations by F. BARNARD. Crown 4to, cloth, 5s.

    WALTER SAVAGE LANDOR: a Biography, 1775-1864.
    With Portrait. A New and Revised Edition. Demy 8vo, 12s.

FORSTER, THE LIFE OF THE RIGHT HON. W. E.
    By T. WEMYSS REID. With Portraits. Fourth Edition. 2 vols. Demy 8vo, 32s.
    FIFTH EDITION, in one volume, with new Portrait. Demy 8vo, 10s. 6d.

FORTESCUE (THE HON. JOHN)—
    RECORDS OF STAG-HUNTING ON EXMOOR. With
    14 full page Illustrations by EDGAR GIBERNE. Large crown 8vo, 16s.

FORTNIGHTLY REVIEW—
    FORTNIGHTLY REVIEW.—First Series, May, 1865, to
    Dec. 1866. 6 vols. Cloth, 13s. each.

    New Series, 1867 to 1872. In Half-yearly Volumes. Cloth,
    13s. each.

    From January, 1873, to the present time, in Half-yearly
    Volumes. Cloth, 16s. each.

    CONTENTS OF FORTNIGHTLY REVIEW. From
    the commencement to end of 1878. Sewed, 2s.

FORTNUM (C. D. E.), F.S.A.—
    MAIOLICA. With numerous Woodcuts. Large crown
    8vo, cloth, 2s. 6d.
    BRONZES. With numerous Woodcuts. Large crown
    8vo, cloth, 2s. 6d.
FOUQUÉ (DE LA MOTTE)—
    UNDINE : a Romance translated from the German. With
    an Introduction by JULIA CARTWRIGHT. Illustrated by HEYWOOD SUMNER.
    Crown 4to. 5s.
FRANCATELLI (C. E.)—
    THE ROYAL CONFECTIONER : English and Foreign.
    A Practical Treatise. With Illustrations. Fifth Edition. Crown 8vo, 5s.
FRANCIS (FRANCIS), JUNR.
    SADDLE AND MOCASSIN. 8vo, 12s.
FRANKS (A. W.)—
    JAPANESE POTTERY. Being a Native Report, with an
    Introduction and Catalogue. With numerous Illustrations and Marks. Large
    crown 8vo, cloth, 2s. 6d.
FROBEL, FRIEDRICH ; a Short Sketch of his Life, including
    Fröbel's Letters from Dresden and Leipzig to his Wife, now first Translated into
    English. By EMILY SHIRREFF. Crown 8vo, 2s.
GALILEO AND HIS JUDGES. By F. R. WEGG-PROSSER.
    Demy 8vo, 5s.
GALLENGA (ANTONIO)—
    ITALY: PRESENT AND FUTURE. 2 vols. Dmy. 8vo, 21s.
    EPISODES OF MY SECOND LIFE. 2 vols. Dmy. 8vo, 28s.
    IBERIAN REMINISCENCES. Fifteen Years' Travelling
    Impressions of Spain and Portugal. With a Map. 2 vols. Demy 8vo, 32s.
GASNAULT (PAUL) and GARNIER (ED.)—
    FRENCH POTTERY. With Illustrations and Marks.
    Large crown 8vo, 3s.
GILLMORE (PARKER)—
    THE HUNTER'S ARCADIA. With numerous Illustra-
    tions. Demy 8vo, 10s. 6d.
GIRL'S LIFE EIGHTY YEARS AGO (A). Selections from
    the Letters of Eliza Southgate Bowne, with an Introduction by Clarence Cook.
    Illustrated with Portraits and Views. Crown 4to, 12s.
GLEICHEN (COUNT), Grenadier Guards—
    WITH THE CAMEL CORPS UP THE NILE. With
    numerous Sketches by the Author. Third Edition. Large crown 8vo, 9s.
GORDON (GENERAL)—
    LETTERS FROM THE CRIMEA, THE DANUBE,
    AND ARMENIA. Edited by DEMETRIUS C. BOULGER. Second Edition.
    Crown 8vo, 5s.
GORST (SIR J. E.), Q.C., M.P.—
    An ELECTION MANUAL. Containing the Parliamentary
    Elections (Corrupt and Illegal Practices) Act, 1883, with Notes. Third Edition.
    Crown 8vo, 1s. 6d.
GOWER (A. R.), Royal School of Mines—
    PRACTICAL METALLURGY. With Illustrations. Crown
    8vo, 3s.
GRAHAM (SIR GERALD), V.C., K.C.B.—
    LAST WORDS WITH GORDON. Crown 8vo, cloth, 1s.

GRESWELL (WILLIAM), M.A., F.R.C.I.—
  OUR SOUTH AFRICAN EMPIRE. With Map. 2 vols.
  Crown 8vo, 21s.

GREVILLE (LADY VIOLET)—
  MONTROSE. With an Introduction by the EARL OF
  ASHBURNHAM. With Portraits. Large crown 8vo, 7s. 6d.

GRIFFIN (SIR LEPEL HENRY), K.C.S.I.—
  THE GREAT REPUBLIC. Second Edition. Crown 8vo,
  4s. 6d.

GRIFFITHS (MAJOR ARTHUR), H.M. Inspector of Prisons—
  FRENCH REVOLUTIONARY GENERALS. Large
  crown 8vo. [In the Press.

  CHRONICLES OF NEWGATE. Illustrated. New
  Edition. Demy 8vo, 16s.

  MEMORIALS OF MILLBANK: or, Chapters in Prison
  History. With Illustrations by R. Goff and Author. New Edition. Demy 8vo,
  12s.

GRIMBLE (AUGUSTUS)—
  DEER-STALKING. A New Edition, revised and enlarged.
  Imperial 4to. With 18 Full-page Illustrations.

HALL (SIDNEY)—
  A TRAVELLING ATLAS OF THE ENGLISH COUN-
  TIES. Fifty Maps, coloured. New Edition, including the Railways, corrected
  up to the present date. Demy 8vo, in roan tuck, 10s. 6d.

HATTON (JOSEPH) and HARVEY (REV. M.)—
  NEWFOUNDLAND. The Oldest British Colony. Its
  History, Past and Present, and its Prospects in the Future. Illustrated from
  Photographs and Sketches specially made for this work. Demy 8vo, 18s.

HAWKINS (FREDERICK)—
  THE FRENCH STAGE IN THE EIGHTEENTH
  CENTURY. With Portraits. 2 vols. Demy 8vo, 30s.

  ANNALS OF THE FRENCH STAGE: FROM ITS
  ORIGIN TO THE DEATH OF RACINE. 4 Portraits. 2 vols. Demy 8vo,
  28s.

HILDEBRAND (HANS), Royal Antiquary of Sweden—
  INDUSTRIAL ARTS OF SCANDINAVIA IN THE
  PAGAN TIME. With numerous Woodcuts. Large crown 8vo, 2s. 6d.

HILL (MISS G.)—
  THE PLEASURES AND PROFITS OF OUR LITTLE
  POULTRY FARM. Small 8vo, 3s.

HOLBEIN—
  TWELVE HEADS AFTER HOLBEIN. Selected from
  Drawings in Her Majesty's Collection at Windsor. Reproduced in Autotype, in
  portfolio. £1 16s.

HOLLINGSHEAD (JOHN)—
  FOOTLIGHTS. Crown 8vo, 7s. 6d.

*HOLMES (GEORGE C. V.), Secretary of the Institution of Naval Architects,
  Whitworth Scholar—*
  MARINE ENGINES AND BOILERS. With Sixty-nine
  Woodcuts. Large crown 8vo, 3s.

HOPE (ANDRÉE)—
 CHRONICLES OF AN OLD INN; or, a Few Words about Gray's Inn. Crown 8vo, 5s.

HOVELACQUE (ABEL)—
 THE SCIENCE OF LANGUAGE: LINGUISTICS, PHILOLOGY, AND ETYMOLOGY. With Maps. Large crown 8vo, cloth, 5s.

HOZIER (H. M.)—
 TURENNE. With Portrait and Two Maps. Large crown 8vo, 4s.

HUEFFER (F.)—
 HALF A CENTURY OF MUSIC IN ENGLAND. 1837—1887. Demy 8vo.

HUMPHRIS (H. D.)—
 PRINCIPLES OF PERSPECTIVE. Illustrated in a Series of Examples. Oblong folio, half-bound, and Text 8vo, cloth, £1 1s.

HUNTLY (MARQUIS OF)—
 TRAVELS, SPORTS, AND POLITICS IN THE EAST OF EUROPE. With Illustrations by the Marchioness of Huntly. Large Crown 8vo, 12s.

 INDUSTRIAL ARTS: Historical Sketches. With numerous Illustrations. Large crown 8vo, 3s.

 INTERNATIONAL POLICY: Essay on the Foreign Relations of England. By FREDERIC HARRISON, PROF. BEESLEY, RICHARD CONGREVE, and others. New Edition. Crown 8vo, 2s. 6d.

 IRELAND IN THE DAYS OF DEAN SWIFT. By J. B. DALY, LL.D. Crown 8vo, 5s.

 IRISH ART OF LACEMAKING, A RENASCENCE OF THE. Illustrated by Photographic Reproductions of Irish Laces, made from new and specially designed Patterns. Introductory Notes and Descriptions. By A. S. C. Demy 8vo, 2s. 6d.

IRON (RALPH), (OLIVE SCHREINER)—
 THE STORY OF AN AFRICAN FARM. New Edition. Crown 8vo, 1s.; in cloth, 2s.

JACKSON (FRANK G.), Master in the Birmingham Municipal School of Art—
 DECORATIVE DESIGN. An Elementary Text Book of Principles and Practice. With numerous Illustrations. Crown 8vo, 7s. 6d.

JAMES (HENRY A.)—
 HANDBOOK TO PERSPECTIVE. Crown 8vo, 2s. 6d.

JARRY (GENERAL)—
 OUTPOST DUTY. Translated, with TREATISES ON MILITARY RECONNAISSANCE AND ON ROAD-MAKING. By Major-Gen. W. C. E. NAPIER. Third Edition. Crown 8vo, 5s.

JEANS (W. T.)—
 CREATORS OF THE AGE OF STEEL. Memoirs of Sir W. Siemens, Sir H. Bessemer, Sir J. Whitworth, Sir J. Brown, and other Inventors. Second Edition. Crown 8vo, 7s. 6d.

JOHNSON (DR. SAMUEL)—
 LIFE AND CONVERSATIONS OF DR. SAMUEL JOHNSON. By A. MAIN. Crown 8vo, 10s. 6d.

JONES (CAPTAIN DOUGLAS), R.A.—
NOTES ON MILITARY LAW. Crown 8vo, 4s.
JONES. HANDBOOK OF THE JONES COLLECTION
IN THE SOUTH KENSINGTON MUSEUM. With Portrait and Woodcuts. Large crown 8vo, 2s. 6d.

KENNARD (EDWARD)—
NORWEGIAN SKETCHES: FISHING IN STRANGE
WATERS. Illustrated with 30 beautiful Sketches printed by The Automatic Engraving Co., and descriptive letterpress. Second Edition. Oblong folio, 21s. A Set of Six Hand-coloured Plates, 21s.; in Oak Frames, 42s.

KENT (CHARLES)—
HUMOUR AND PATHOS OF CHARLES DICKENS.
Crown 8vo, 6s.

KLACZKO (M. JULIAN)—
TWO CHANCELLORS: PRINCE GORTCHAKOF AND
PRINCE BISMARCK. Translated by MRS. TAIT. New and cheaper Edition, 6s.

KNOLLYS (MAJOR HENRY), R.A.—
SKETCHES OF LIFE IN JAPAN. With Illustrations.
Large crown 8vo 12s.

LACEMAKING, A RENASCENCE OF THE IRISH
ART OF. Illustrated by Photographic Reproductions of Irish Laces, made from new and specially designed patterns. Demy 8vo, 2s. 6d.

LACORDAIRE'S JESUS CHRIST; GOD; AND GOD AND
MAN. Conferences delivered at Notre Dame in Paris. New Edition. Crown 8vo, 6s.

LAING (S.)—
MODERN SCIENCE AND MODERN THOUGHT.
With a Supplementary Chapter on Gladstone's "Dawn of Creation" and Drummond's "Natural Law in the Spiritual World." Sixth Thousand. Demy 8vo, 3s. 6d.

LAVELEYE (ÉMILE DE)—
THE ELEMENTS OF POLITICAL ECONOMY.
Translated by W. POLLARD, B.A., St. John's College, Oxford. Crown 8vo, 6s.

LANDOR (W. S.)—
LIFE AND WORKS. 8 vols.
VOL. 1. WALTER SAVAGE LANDOR. A Biography in Eight Books. By JOHN FORSTER. Demy 8vo, 12s.
VOL. 2. Out of print.
VOL. 3. CONVERSATIONS OF SOVEREIGNS AND STATESMEN, AND FIVE DIALOGUES OF BOCCACCIO AND PETRARCA. Demy 8vo, 14s.
VOL. 4. DIALOGUES OF LITERARY MEN. Demy 8vo, 14s.
VOL. 5. DIALOGUES OF LITERARY MEN (*continued*). FAMOUS WOMEN. LETTERS OF PERICLES AND ASPASIA, And Minor Prose Pieces. Demy 8vo, 14s.
VOL. 6. MISCELLANEOUS CONVERSATIONS. Demy 8vo, 14s.
VOL. 7. GEBIR, ACTS AND SCENES AND HELLENICS. Poems. Demy 8vo, 14s.
VOL. 8. MISCELLANEOUS POEMS AND CRITICISMS ON THEOCRITUS, CATULLUS, AND PETRARCH. Demy 8vo, 14s.

LE CONTE (JOSEPH), Professor of Geology and Natural History in the University of California—
EVOLUTION AND ITS RELATIONS TO RELIGIOUS
THOUGHT. Crown 8vo, 6s.

LEFÈVRE (ANDRÉ)—
> PHILOSOPHY, Historical and Critical. Translated, with an Introduction, by A. W. KEANE, B.A. Large crown 8vo, 7s. 6d.

LESLIE (R. C.)—
> LIFE ABOARD A BRITISH PRIVATEER IN THE TIME OF QUEEN ANNE. Being the Journals of Captain Woodes Rogers, Master Mariner. With Notes and Illustrations by ROBERT C. LESLIE. Large crown 8vo, 9s.
>
> A SEA PAINTER'S LOG. With 12 Full-page Illustrations by the Author. Large crown 8vo, 12s.

LETOURNEAU (DR. CHARLES)—
> SOCIOLOGY. Based upon Ethnology. Large crown 8vo, 10s.
>
> BIOLOGY. Translated by WILLIAM MACCALL. With Illustrations. Large crown 8vo, 6s.

LILLY (W. S.)—
> CHAPTERS ON EUROPEAN HISTORY. With an Introductory Dialogue on the Philosophy of History. 2 vols. Demy 8vo, 21s.
>
> ANCIENT RELIGION AND MODERN THOUGHT. Third Edition, revised, with additions. Demy 8vo, 12s.

LITTLE (THE REV. CANON KNOX)—
> THE CHILD OF STAFFERTON: A Chapter from a Family Chronicle. Tenth Thousand. Crown 8vo, 2s. 6d.
>
> THE BROKEN VOW. A Story of Here and Hereafter. Tenth Thousand. Crown 8vo, 2s. 6d.

LLOYD (COLONEL E.M.), R.E., late Professor of Fortification at the Royal Military Academy, Woolwich—
> VAUBAN, MONTALEMBERT, CARNOT: ENGINEER STUDIES. With Portraits. Crown 8vo, 5s.

LONG (JAMES)—
> DAIRY FARMING. To which is added a Description of the Chief Continental Systems. With numerous Illustrations. Crown 8vo, 9s.

LOW (C. R.)—
> SOLDIERS OF THE VICTORIAN AGE. 2 vols. Demy 8vo, £1 10s.

LOW (WILLIAM)—
> TABLE DECORATION. With 19 Full Illustrations. Demy 8vo, 6s.

LYTTON (ROBERT, EARL)—
> POETICAL WORKS—
>> FABLES IN SONG. 2 vols. Fcap. 8vo, 12s.
>> THE WANDERER. Fcap. 8vo, 6s.
>> POEMS, HISTORICAL AND CHARACTERISTIC. Fcap. 6s.

MACDONALD (FREDERIKA)—
   PUCK AND PEARL: THE WANDERINGS AND WONDER-
   INGS OF TWO ENGLISH CHILDREN IN INDIA. By FREDERIKA MACDONALD.
   With Illustrations by MRS. IRVING GRAHAM. Second Edition. Crown 8vo, 5s.

MALLESON (COL. G. B.), C.S.I.—
   PRINCE EUGENE OF SAVOY. With Portrait and
   Maps. Large crown 8vo, 6s.
   LOUDON. A Sketch of the Military Life of Gideon
   Ernest, Freiherr von Loudon, sometime Generalissimo of the Austrian Forces.
   With Portrait and Maps. Large crown 8vo, 4s.

MALLET (ROBERT)—
   PRACTICAL MANUAL OF CHEMICAL ASSAYING,
   as applied to the Manufacture of Iron. By L. L. DE KONINCK and E. DIETZ.
   Edited, with notes, by ROBERT MALLET. Post 8vo, cloth, 6s.

MASKELL (ALFRED)—
   RUSSIAN ART AND ART OBJECTS IN RUSSIA.
   A Handbook to the Reproduction of Goldsmiths' Work and other Art Treasures.
   With Illustrations. Large crown 8vo, 4s. 6d.

MASKELL (WILLIAM)—
   IVORIES: ANCIENT AND MEDIÆVAL. With nume-
   rous Woodcuts. Large crown 8vo, cloth, 2s. 6d.
   HANDBOOK TO THE DYCE AND FORSTER COL-
   LECTIONS. With Illustrations. Large crown 8vo, cloth, 2s. 6d.

MAUDSLAY (ATHOL)—
   HIGHWAYS AND HORSES. With numerous Illustrations. Demy 8vo, 21s.

MECHELIN (SENATOR L.)—
   FINLAND AND ITS PUBLIC LAW. Translated by
   CHARLES J. COOKE, British Vice-Consul at Helsingfors. Crown 8vo, 2s. 6d.

GEORGE MEREDITH'S WORKS.
A New and Uniform Edition. Crown 8vo, 3s. 6d. each.

DIANA OF THE CROSSWAYS.
EVAN HARRINGTON.
THE ORDEAL OF RICHARD FEVEREL.
THE ADVENTURES OF HARRY RICHMOND.
SANDRA BELLONI.
VITTORIA.
RHODA FLEMING.
BEAUCHAMP'S CAREER.
THE EGOIST.
THE SHAVING OF SHAGPAT; AND FARINA.

C

MERIVALE (HERMAN CHARLES)—
BINKO'S BLUES. A Tale for Children of all Growths. Illustrated by Edgar Giberne. Small crown 8vo, 5s.

THE WHITE PILGRIM, and other Poems. Crown 8vo, 9s.

MOLESWORTH (W. NASSAU)—
HISTORY OF ENGLAND FROM THE YEAR 1830 TO THE RESIGNATION OF THE GLADSTONE MINISTRY, 1874. Twelfth Thousand. 3 vols. Crown 8vo, 18s.

ABRIDGED EDITION. Large crown, 7s. 6d.

MOLTKE (FIELD-MARSHAL COUNT VON)—
POLAND: AN HISTORICAL SKETCH. An Authorised Translation, with Biographical Notice by E. S. Buchheim. Crown 8vo, 4s. 6d.

MORLEY (THE RIGHT HON. JOHN), M.P.—
RICHARD COBDEN'S LIFE AND CORRESPONDENCE. Crown 8vo, with Portrait, 7s. 6d.

Popular Edition. With Portrait. 4to, sewed, 1s. Cloth, 2s.

MUNTZ (EUGENE)—
RAPHAEL: his Life, Works, and Times. Illustrated with about 200 Engravings. A new Edition, revised from the Second French Edition by W. Armstrong, B.A. Oxon. Imperial 8vo, 25s.

MURRAY (ANDREW), F.L.S.—
ECONOMIC ENTOMOLOGY. Aptera. With numerous Illustrations. Large crown 8vo, 7s. 6d.

NAPIER (MAJ.-GEN. W. C. E.)—
TRANSLATION OF GEN. JARRY'S OUTPOST DUTY. With TREATISES ON MILITARY RECONNAISSANCE AND ON ROAD-MAKING. Third Edition. Crown 8vo, 5s.

NAPOLEON. A Selection from the Letters and Despatches of the First Napoleon. With Explanatory Notes by Captain the Hon. D. Bingham. 3 vols. Demy 8vo, £2 2s.

NECKER (MADAME)—
THE SALON OF MADAME NECKER. By Vicomte d'Haussonville. 2 vols. Crown 8vo, 18s.

NESBITT (ALEXANDER)—
GLASS. With numerous Woodcuts. Large crown 8vo, cloth, 2s. 6d.

NEVINSON (HENRY)—
A SKETCH OF HERDER AND HIS TIMES. With a Portrait. Demy 8vo, 14s.

NEWTON (R. TULLEY), F.G.S.—

**THE TYPICAL PARTS IN THE SKELETONS OF
A CAT, DUCK, AND CODFISH,** being a Catalogue with Comparative
Description arranged in a Tabular form. Demy 8vo, cloth, 3s.

NILSEN (CAPTAIN)—

**LEAVES FROM THE LOG OF THE "HOMEWARD
BOUND";** or, Eleven Months at Sea in an Open Boat. Crown 8vo, 1s.

NORMAN (C. B.)—

TONKIN; OR, FRANCE IN THE FAR EAST. With
Maps. Demy 8vo, 14s.

O'GRADY (STANDISH)—

TORYISM AND THE TORY DEMOCRACY. Crown
8vo, 5s.

OLIVER (PROFESSOR), F.R.S., &c.—

**ILLUSTRATIONS OF THE PRINCIPAL NATURAL
ORDERS OF THE VEGETABLE KINGDOM,** PREPARED FOR THE
SCIENCE AND ART DEPARTMENT, SOUTH KENSINGTON. With
109 Plates. Oblong 8vo, plain, 16s.; coloured, £1 6s.

OXENHAM (REV. H. N.)—

MEMOIR OF LIEUTENANT RUDOLPH DE LISLE,
R.N., OF THE NAVAL BRIGADE. Third Edition, with Illustrations.
Crown 8vo, 7s. 6d.

SHORT STUDIES, ETHICAL AND RELIGIOUS.
Demy 8vo, 12s.

**SHORT STUDIES IN ECCLESIASTICAL HISTORY
AND BIOGRAPHY.** Demy 8vo, 12s.

PAYTON (E. W.)—

ROUND ABOUT NEW ZEALAND. Being Notes from
a Journal of Three Years' Wandering in the Antipodes. With Twenty Original
Illustrations by the Author. Large crown 8vo. 12s.

PERROT (GEORGES) and CHIPIEZ (CHARLES)—

**A HISTORY OF ANCIENT ART IN PHŒNICIA
AND ITS DEPENDENCIES.** Translated from the French by WALTER
ARMSTRONG, B.A. Oxon. Containing 644 Illustrations in the text, and 10 Steel
and Coloured Plates. 2 vols. Imperial 8vo, 42s.

A HISTORY OF ART IN CHALDÆA AND ASSYRIA.
Translated by WALTER ARMSTRONG, B.A. Oxon. With 452 Illustrations. 2 vols.
Imperial 8vo, 42s.

A HISTORY OF ART IN ANCIENT EGYPT. Translated from the French by W. ARMSTRONG, B.A. Oxon. With over 600 Illustrations. 2 vols. Imperial 8vo, 42s.

PETERBOROUGH (THE EARL OF)—

THE EARL OF PETERBOROUGH AND MON-
MOUTH (Charles Mordaunt): A Memoir. By Colonel FRANK RUSSELL, Royal
Dragoons. With Illustrations. 2 vols. demy 8vo. 32s.

PHŒNICIAN ART—

A HISTORY OF ANCIENT ART IN PHŒNICIA
AND ITS DEPENDENCIES. By GEORGES PERROT and CHARLES CHIPIEZ.
Translated from the French by WALTER ARMSTRONG, B.A. Oxon. Containing
644 Illustrations in the text, and 10 Steel and Coloured Plates. 2 vols. Imperial
8vo, 42s.

PITT TAYLOR (FRANK)—

THE CANTERBURY TALES. Selections from the Tales
of GEOFFREY CHAUCER rendered into Modern English, with close adherence
to the language of the Poet. With Frontispiece. Crown 8vo, 6s.

POLLEN (J. H.)—

GOLD AND SILVER SMITH'S WORK. With numerous Woodcuts. Large crown 8vo, cloth, 2s. 6d.

ANCIENT AND MODERN FURNITURE AND
WOODWORK. With numerous Woodcuts. Large crown 8vo, cloth, 2s. 6d.

POOLE (STANLEY LANE), B.A., M.R.A.S.—

THE ART OF THE SARACENS IN EGYPT. Published for the Committee of Council on Education. With 108 Woodcuts. Large
crown 8vo, 4s.

POYNTER (E. J.), R.A.—

TEN LECTURES ON ART. Third Edition. Large
crown 8vo, 9s.

PRINSEP (VAL), A.R.A.—

IMPERIAL INDIA. Containing numerous Illustrations
and Maps. Second Edition. Demy 8vo, £1 1s.

RADICAL PROGRAMME, THE. From the *Fortnightly*
Review, with additions. With a Preface by the RIGHT HON. J. CHAMBERLAIN,
M.P. Thirteenth Thousand. Crown 8vo, 2s. 6d.

RAE (W. FRASER)—

AUSTRIAN HEALTH RESORTS: and the Bitter Waters
of Hungary. Crown 8vo, 5s.

RAMSDEN (LADY GWENDOLEN)—

A BIRTHDAY BOOK. Illustrated. Containing 46 Illustrations from Original Drawings, and numerous other Illustrations. Royal 8vo, 21s.

RAPHAEL: his Life, Works, and Times. By EUGENE MUNTZ.
Illustrated with about 200 Engravings. A New Edition, revised from the Second French Edition. By W. ARMSTRONG, B.A. Imperial 8vo, 25s.

REDGRAVE (GILBERT)—
OUTLINES OF HISTORIC ORNAMENT. Translated from the German. Edited by GILBERT REDGRAVE. With numerous Illustrations. Crown 8vo, 4s.

REDGRAVE (GILBERT R.)—
MANUAL OF DESIGN, compiled from the Writings and Addresses of RICHARD REDGRAVE, R.A. With Woodcuts. Large crown 8vo, cloth, 2s. 6d.

REDGRAVE (RICHARD)—
ELEMENTARY MANUAL OF COLOUR, with a Catechism on Colour. 24mo, cloth, 9d.

REDGRAVE (SAMUEL)—
A DESCRIPTIVE CATALOGUE OF THE HISTORICAL COLLECTION OF WATER-COLOUR PAINTINGS IN THE SOUTH KENSINGTON MUSEUM. With numerous Chromo-lithographs and other Illustrations. Royal 8vo, £1 1s.

REID (T. WEMYSS)—
THE LIFE OF THE RIGHT HON. W. E. FORSTER.
With Portraits. Fourth Edition. 2 vols. Demy 8vo, 32s.
FIFTH EDITION, in one volume, with new Portrait. Demy 8vo, 10s. 6d.

RENAN (ERNEST)—
HISTORY OF THE PEOPLE OF ISRAEL TILL THE TIME OF KING DAVID. Demy 8vo, 14s.

HISTORY OF THE PEOPLE OF ISRAEL. From the Reign of David up to the Capture of Samaria. Second Division. Demy 8vo, 14s.

RECOLLECTIONS OF MY YOUTH. Translated from the original French, and revised by MADAME RENAN. Crown 8vo, 8s.

REYNARDSON (C. T. S. BIRCH)—
SPORTS AND ANECDOTES OF BYGONE DAYS in England, Scotland, Ireland, Italy, and the Sunny South. With numerous Illustrations in Colour. Second Edition. Large crown 8vo, 12s.

DOWN THE ROAD: Reminiscences of a Gentleman Coachman. With Coloured Illustrations. Large crown 8vo, 12s.

RIAÑO (JUAN F.)—
THE INDUSTRIAL ARTS IN SPAIN. With numerous Woodcuts. Large crown 8vo, cloth, 4s.

RIBTON-TURNER (C. J.)—
   A HISTORY OF VAGRANTS AND VAGRANCY AND
   BEGGARS AND BEGGING. With Illustrations. Demy 8vo, 21s.

RODINSON (JAMES F.)—
   BRITISH BEE FARMING. Its Profits and Pleasures.
   Large crown 8vo, 5s.

ROBINSON (J. C.)—
   ITALIAN SCULPTURE OF THE MIDDLE AGES
   AND PERIOD OF THE REVIVAL OF ART. With 20 Engravings. Royal
   8vo, cloth, 7s. 6d.

ROBSON (GEORGE)—
   ELEMENTARY BUILDING CONSTRUCTION. Illustrated by a Design for an Entrance Lodge and Gate. 15 Plates. Oblong folio, sewed, 8s.

ROBSON (REV. J. H.), M.A., LL.M.—
   AN ELEMENTARY TREATISE ON ALGEBRA.
   Post 8vo, 6s.

ROCK (THE VERY REV. CANON), D.D.—
   TEXTILE FABRICS. With numerous Woodcuts. Large
   crown 8vo, cloth, 2s. 6d.

ROGERS (CAPTAIN WOODES), Master Mariner—
   LIFE ABOARD A BRITISH PRIVATEER IN THE
   TIME OF QUEEN ANNE. Being the Journals of Captain Woodes Rogers, Master Mariner. With Notes and Illustrations by ROBERT C. LESLIE, Author of "A Sea Painter's Log." Large crown 8vo, 9s.

ROOSE (ROBSON), M.D., F.C.S.—
   THE WEAR AND TEAR OF LONDON LIFE.
   Second Edition. Crown 8vo, sewed, 1s.

   INFECTION AND DISINFECTION. Crown 8vo, sewed, 6d.

ROLAND (ARTHUR)—
   FARMING FOR PLEASURE AND PROFIT. Edited
   by WILLIAM ABLETT. 8 vols. Crown 8vo, 5s. each.
   DAIRY-FARMING, MANAGEMENT OF COWS, &c.
   POULTRY-KEEPING.
   TREE-PLANTING, FOR ORNAMENTATION OR PROFIT.
   STOCK-KEEPING AND CATTLE-REARING.
   DRAINAGE OF LAND, IRRIGATION, MANURES, &c.
   ROOT-GROWING, HOPS, &c.
   MANAGEMENT OF GRASS LANDS, LAYING DOWN GRASS, ARTIFICIAL GRASSES, &c.
   MARKET GARDENING, HUSBANDRY FOR FARMERS AND GENERAL CULTIVATORS.

RUSDEN (G. W.), for many years Clerk of the Parliament in Victoria—
A HISTORY OF AUSTRALIA. With a Coloured Map.
3 vols. Demy 8vo, 50s.

RUSSELL (COLONEL FRANK), Royal Dragoons—
THE EARL OF PETERBOROUGH AND MONMOUTH (Charles Mordaunt): A Memoir. With Illustrations. 2 vols. demy 8vo, 32s.

"RUSSIA'S HOPE," THE; OR, BRITANNIA NO LONGER RULES THE WAVES. Showing how the Muscovite Bear got at the British Whale. Translated from the original Russian by CHARLES JAMES COOKE. Crown 8vo, 1s.

SCIENCE AND ART: a Journal for Teachers and Scholars. Issued monthly. 3d. See page 39.

SCOTT (MAJOR-GENERAL A. DE C.), late Royal Engineers—
LONDON WATER: a Review of the Present Condition and Suggested Improvements of the Metropolitan Water Supply. Crown 8vo, sewed, 2s.

SCOTT (LEADER)—
THE RENAISSANCE OF ART IN ITALY: an Illustrated Sketch. With upwards of 200 Illustrations. Medium quarto, 18s.

SCOTT-STEVENSON (MRS.)—
ON SUMMER SEAS. Including the Mediterranean, the Ægean, the Ionian, and the Euxine, and a voyage down the Danube. With a Map. Demy 8vo, 16s.

OUR HOME IN CYPRUS. With a Map and Illustrations. Third Edition. Demy 8vo, 14s.

OUR RIDE THROUGH ASIA MINOR. With Map. Demy 8vo, 18s.

SEEMAN (O.)—
THE MYTHOLOGY OF GREECE AND ROME, with Special Reference to its Use in Art. From the German. Edited by G. H. BIANCHI. 64 Illustrations. New Edition. Crown 8vo, 5s.

SETON-KARR (H. W.), F.R.G.S., etc.—
TEN YEARS' WILD SPORTS IN FOREIGN LANDS; or, Travels in the Eighties. Demy 8vo.

SHEPHERD (MAJOR), R.E.—
PRAIRIE EXPERIENCES IN HANDLING CATTLE AND SHEEP. With Illustrations and Map. Demy 8vo, 10s. 6d.

SHIRREFF (EMILY)—
A SHORT SKETCH OF THE LIFE OF FRIEDRICH FROBEL; a New Edition, including Fröbel's Letters from Dresden and Leipzig to his Wife, now first Translated into English. Crown 8vo, 2s.

HOME EDUCATION IN RELATION TO THE KINDERGARTEN. Two Lectures. Crown 8vo, 1s. 6d.

SHORE (ARABELLA)—
DANTE FOR BEGINNERS: a Sketch of the "Divina Commedia." With Translations, Biographical and Critical Notices, and Illustrations. With Portrait. Crown 8vo, 6s.

SIMMONDS (T. L.)—
    ANIMAL PRODUCTS: their Preparation, Commercial Uses, and Value. With numerous Illustrations. Large crown 8vo, 7s. 6d.

    SINGER'S STORY, A. Related by the Author of "Flitters, Tatters, and the Counsellor." Crown 8vo, sewed, 1s.

SINNETT (A. P.)—
    ESOTERIC BUDDHISM. Annotated and enlarged by the Author. Sixth and cheaper Edition. Crown 8vo, 4s.

    KARMA. A Novel. New Edition. Crown 8vo, 3s. 6d.

SINNETT (MRS.)—
    THE PURPOSE OF THEOSOPHY. Crown 8vo, 3s.

SMITH (ALEXANDER SKENE)—
    HOLIDAY RECREATIONS, AND OTHER POEMS. With a Preface by Rev. Principal Cairns, D.D. Crown 8vo, 5s.

SMITH (MAJOR R. MURDOCK), R.E.—
    PERSIAN ART. With Map and Woodcuts. Second Edition. Large crown 8vo, 2s.

STOKES (MARGARET)—
    EARLY CHRISTIAN ART IN IRELAND. With 106 Woodcuts. Demy 8vo, 7s. 6d.

STORY (W. W.)—
    ROBA DI ROMA. Seventh Edition, with Additions and Portrait. Crown 8vo, cloth, 10s. 6d.

    CASTLE ST. ANGELO. With Illustrations. Crown 8vo, 10s. 6d.

    A SUBURB OF YEDO. By T. A. P. With Illustrations. Crown 8vo.

SUTCLIFFE (JOHN)—
    THE SCULPTOR AND ART STUDENT'S GUIDE to the Proportions of the Human Form, with Measurements in feet and inches of Full-Grown Figures of Both Sexes and of Various Ages. By Dr. G. Schadow, Member of the Academies, Stockholm, Dresden, Rome, &c. &c. Translated by J. J. Wright. Plates reproduced by J. Sutcliffe. Oblong folio, 31s. 6d.

TAINE (H. A.)—
    NOTES ON ENGLAND. Translated, with Introduction, by W. Fraser Rae. Eighth Edition. With Portrait. Crown 8vo, 5s.

TANNER (PROFESSOR), F.C.S.—
    HOLT CASTLE; or, Threefold Interest in Land. Crown 8vo, 4s. 6d.

    JACK'S EDUCATION; OR, HOW HE LEARNT FARMING. Second Edition. Crown 8vo, 4s. 6d.

TEMPLE (SIR RICHARD), BART., M.P., G.C.S.I.—
    COSMOPOLITAN ESSAYS. With Maps. Demy 8vo, 16s.

THRUPP (GEORGE A.) and FARR (WILLIAM)—
    COACH TRIMMING. With 60 Illustrations. Crown 8vo, 2s. 6d.

TOPINARD (DR. PAUL)—
    ANTHROPOLOGY. With a Preface by Professor PAUL BROCA. With numerous Illustrations. Large crown 8vo, 7s. 6d.

TOVEY (LIEUT.-COL., R.E.)—
    MARTIAL LAW AND CUSTOM OF WAR; or, Military Law and Jurisdiction in Troublous Times. Crown 8vo, 6s.

TRAHERNE (MAJOR)—
    THE HABITS OF THE SALMON. Crown 8vo.

TRAILL (H. D.)—
    THE NEW LUCIAN. Being a Series of Dialogues of the Dead. Demy 8vo, 12s.

TROLLOPE (ANTHONY)—
    THE CHRONICLES OF BARSETSHIRE. A Uniform Edition, in 8 vols., large crown 8vo, handsomely printed, each vol. containing Frontispiece. 6s. each.

    THE WARDEN and BARCHESTER TOWERS. 2 vols.
    DR. THORNE.
    FRAMLEY PARSONAGE.
    THE SMALL HOUSE AT ALLINGTON. 2 vols.
    LAST CHRONICLE OF BARSET. 2 vols.

    LIFE OF CICERO. 2 vols. 8vo. £1 4s.

VERON (EUGENE)—
    ÆSTHETICS. Translated by W. H. ARMSTRONG. Large crown 8vo, 7s. 6d.

VERSCHOYLE (REV. J.), M.A.—
    THE HISTORY OF ANCIENT CIVILISATION. A Handbook based upon M. Gustave Ducoudray's "Histoire Sommaire de la Civilisation." Edited by REV. J. VERSCHOYLE, M.A. With Illustrations. Large crown 8vo, 6s.

WALE (REV. HENRY JOHN), M.A.—
    MY GRANDFATHER'S POCKET BOOK, from 1701 to 1796. Author of "Sword and Surplice." Demy 8vo, 12s.

WALFORD (MAJOR), R.A.—
    PARLIAMENTARY GENERALS OF THE GREAT CIVIL WAR. With Maps. Large crown 8vo, 4s.

WALKER (MRS.)—
    UNTRODDEN PATHS IN ROUMANIA. With 77 Illustrations. Demy 8vo, 10s. 6d.

    EASTERN LIFE AND SCENERY, with Excursions to Asia Minor, Mitylene, Crete, and Roumania. 2 vols., with Frontispiece to each vol. Crown 8vo, 21s.

WARING (CHARLES)—
    STATE PURCHASE OF RAILWAYS. Demy 8vo, 5s.

WATSON (WILLIAM)—
    LIFE IN THE CONFEDERATE ARMY: being the Observations and Experiences of an Alien in the South during the American Civil War. Crown 8vo, 6s.

WEGG-PROSSER (F. R.)—
    GALILEO AND HIS JUDGES. Demy 8vo, 5s.

WHITE (WALTER)—
    A MONTH IN YORKSHIRE. With a Map. Fifth Edition. Post 8vo, 4s.

    A LONDONER'S WALK TO THE LAND'S END, AND A TRIP TO THE SCILLY ISLES. With 4 Maps. Third Edition. Post 8vo, 4s.

WILL-O'-THE-WISPS, THE. Translated from the German of Marie Petersen by CHARLOTTE J. HART. With Illustrations. Crown 8vo, 7s. 6d.

WORKING MAN'S PHILOSOPHY, A. By "ONE OF THE CROWD." Crown 8vo, 3s.

WORNUM (R. N.)—
    ANALYSIS OF ORNAMENT: THE CHARACTERISTICS OF STYLES. An Introduction to the History of Ornamental Art. With many Illustrations. Ninth Edition. Royal 8vo, cloth 8s.

WRIGHTSON (PROF. JOHN), M.R.A.C., F.C.S., &c.; Examiner in Agriculture to the Science and Art Department; Professor of Agriculture in the Normal School of Science and Royal School of Mines; President of the College of Agriculture, Downton, near Salisbury; late Commissioner for the Royal Agricultural Society of England, &c., &c.

    PRINCIPLES OF AGRICULTURAL PRACTICE AS AN INSTRUCTIONAL SUBJECT. With Geological Map. Crown 8vo, 5s.

    FALLOW AND FODDER CROPS. [*In the Press.*

WORSAAE (J. J. A.)—
    INDUSTRIAL ARTS OF DENMARK, FROM THE EARLIEST TIMES TO THE DANISH CONQUEST OF ENGLAND. With Maps and Woodcuts. Large crown 8vo, 3s. 6d.

YEO (DR. J. BURNEY)—
    CLIMATE AND HEALTH RESORTS. New Edition. Crown 8vo, 10s. 6d.

YOUNGE (C. D.)—
    PARALLEL LIVES OF ANCIENT AND MODERN HEROES. New Edition. 12mo, cloth, 4s. 6d.

WINDT (H. DE)—
    FROM CALAIS TO PEKIN BY LAND. With Numerous Illustrations by the Author. Demy 8vo.

YOUNG OFFICER'S "DON'T"; or, Hints to Youngsters on Joining. 32mo 1s.

SOUTH KENSINGTON MUSEUM SCIENCE AND ART HANDBOOKS.

Handsomely printed in large crown 8vo.

Published for the Committee of the Council on Education.

MARINE ENGINES AND BOILERS. By GEORGE C. V. HOLMES, Secretary of the Institution of Naval Architects, Whitworth Scholar. With Sixty-nine Woodcuts. Large crown 8vo, 3s.

EARLY CHRISTIAN ART IN IRELAND. By MARGARET STOKES. With 106 Woodcuts. Crown 8vo, 4s.
A Library Edition, demy 8vo, 7s. 6d.

FOOD GRAINS OF INDIA. By PROF. A. H. CHURCH, M.A., F.C.S., F.I.C. With Numerous Woodcuts. Small 4to, 6s.

THE ART OF THE SARACENS IN EGYPT. By STANLEY LANE POOLE, B.A., M.A.R.S With 108 Woodcuts. Crown 8vo, 4s.

ENGLISH PORCELAIN: A Handbook to the China made in England during the 18th Century, as illustrated by Specimens chiefly in the National Collections. By PROF. A. H. CHURCH, M.A. With numerous Woodcuts. 3s.

RUSSIAN ART AND ART OBJECTS IN RUSSIA: A Handbook to the reproduction of Goldsmiths' work and other Art Treasures from that country in the South Kensington Museum. By ALFRED MASKELL. With Illustrations. 4s. 6d.

FRENCH POTTERY. By PAUL GASNAULT and EDOUARD GARNIER. With Illustrations and Marks. 3s.

ENGLISH EARTHENWARE: A Handbook to the Wares made in England during the 17th and 18th Centuries, as illustrated by Specimens in the National Collection. By PROF. A. H. CHURCH, M.A. With numerous Woodcuts. 3s.

INDUSTRIAL ARTS OF DENMARK. From the Earliest Times to the Danish Conquest of England. By J. J. A. WORSAAE, Hon. F.S.A., &c. &c. With Map and Woodcuts. 3s. 6d.

INDUSTRIAL ARTS OF SCANDINAVIA IN THE PAGAN TIME. By HANS HILDEBRAND, Royal Antiquary of Sweden. With numerous Woodcuts. 2s. 6d.

PRECIOUS STONES: Considered in their Scientific and Artistic relations, with a Catalogue of the Townsend Collection of Gems in the South Kensington Museum. By PROF. A. H. CHURCH, M.A. With a Coloured Plate and Woodcuts. 2s. 6d.

INDUSTRIAL ARTS OF INDIA. By Sir GEORGE C. M. BIRDWOOD, C.S.I., &c. With Map and Woodcuts. Demy 8vo, 14s.

HANDBOOK TO THE DYCE AND FORSTER COLLECTIONS in the South Kensington Museum. With Portraits and Facsimiles. 2s. 6d.

INDUSTRIAL ARTS IN SPAIN. By JUAN F. RIAÑO. With numerous Woodcuts. 4s.

GLASS. By ALEXANDER NESBITT. With numerous Woodcuts. 2s. 6d.

GOLD AND SILVER SMITHS' WORK. By JOHN HUNGERFORD POLLEN, M.A. With numerous Woodcuts. 2s. 6d.

TAPESTRY. By ALFRED DE CHAMPEAUX. With Woodcuts. 2s. 6d.

BRONZES. By C. DRURY E. FORTNUM, F.S.A. With numerous Woodcuts. 2s. 6d.

SOUTH KENSINGTON MUSEUM SCIENCE & ART HANDBOOKS—*Continued.*

PLAIN WORDS ABOUT WATER. By A. H. CHURCH, M.A.
Oxon. With Illustrations. Sewed, 6d.

ANIMAL PRODUCTS: their Preparation, Commercial Uses,
and Value. By T. L. SIMMONDS. With Illustrations. 7s. 6d.

FOOD: Some Account of its Sources, Constituents, and Uses.
By PROFESSOR A. H. CHURCH, M.A. Oxon. Sixth Thousand. 3s.

ECONOMIC ENTOMOLOGY. By ANDREW MURRAY, F.L.S.
APTERA. With Illustrations. 7s. 6d.

JAPANESE POTTERY. Being a Native Report. With an
Introduction and Catalogue by A. W. FRANKS, M.A., F.R.S., F.S.A. With Illustrations and Marks. 2s. 6d.

HANDBOOK TO THE SPECIAL LOAN COLLECTION
of Scientific Apparatus. 3s.

INDUSTRIAL ARTS: Historical Sketches. With Numerous
Illustrations. 3s.

TEXTILE FABRICS. By the Very Rev. DANIEL ROCK, D.D.
With numerous Woodcuts. 2s. 6d.

JONES COLLECTION IN THE SOUTH KENSINGTON
MUSEUM. With Portrait and Woodcuts. 2s. 6d.

COLLEGE AND CORPORATION PLATE. A Handbook
to the Reproductions of Silver Plate in the South Kensigton Museum from Celebrated English Collections. By WILFRED JOSEPH CRIPPS, M.A., F.S.A. With Illustrations. 2s. 6d.

IVORIES: ANCIENT AND MEDIÆVAL. By WILLIAM
MASKELL. With numerous Woodcuts. 2s. 6d.

ANCIENT AND MODERN FURNITURE AND WOOD-
WORK. By JOHN HUNGERFORD POLLEN, M.A. With numerous Woodcuts. 2s. 6d.

MAIOLICA. By C. DRURY E. FORTNUM, F.S.A. With
numerous Woodcuts. 2s. 6d.

THE CHEMISTRY OF FOODS. With Microscopic Illus-
trations. By JAMES BELL, Ph.D., &c., Principal of the Somerset House Laboratory. Part I.—Tea, Coffee, Cocoa, Sugar, &c. 2s. 6d.
Part II.—Milk, Butter, Cheese, Cereals, Prepared Starches, &c. 3s.

MUSICAL INSTRUMENTS. By CARL ENGEL. With numerous Woodcuts. 2s. 6d

MANUAL OF DESIGN, compiled from the Writings and
Addresses of RICHARD REDGRAVE, R.A. By GILBERT R. REDGRAVE. With Woodcuts. 2s. 6d.

PERSIAN ART. By MAJOR R. MURDOCK SMITH, R.E. With
Map and Woodcuts. Second Edition, enlarged. 2s.

CARLYLE'S (THOMAS) WORKS.

THE ASHBURTON EDITION.

An entirely New Edition, handsomely printed, containing all the Portraits and Illustrations, in Seventeen Volumes, demy 8vo, 8s. each.

THE FRENCH REVOLUTION AND PAST AND PRESENT. 2 vols.
SARTOR RESARTUS; HEROES AND HERO WORSHIP. 1 vol.
LIFE OF JOHN STERLING—LIFE OF SCHILLER. 1 vol.
LATTER-DAY PAMPHLETS—EARLY KINGS OF NORWAY—ESSAY ON THE PORTRAIT OF JOHN KNOX. 1 vol.
LETTERS AND SPEECHES OF OLIVER CROMWELL. 3 vols.
HISTORY OF FREDERICK THE GREAT. 6 vols.
CRITICAL AND MISCELLANEOUS ESSAYS. 3 vols.

LIBRARY EDITION COMPLETE.

Handsomely printed in 34 vols., demy 8vo, cloth, £15 3s.

SARTOR RESARTUS. With a Portrait, 7s. 6d.

THE FRENCH REVOLUTION. A History. 3 vols., each 9s.

LIFE OF FREDERICK SCHILLER AND EXAMINATION OF HIS WORKS. With Supplement of 1872. Portrait and Plates, 9s.

CRITICAL AND MISCELLANEOUS ESSAYS. With Portrait. 6 vols., each 9s.

ON HEROES, HERO WORSHIP, AND THE HEROIC IN HISTORY. 7s. 6d.

PAST AND PRESENT. 9s.

OLIVER CROMWELL'S LETTERS AND SPEECHES. With Portraits. 5 vols., each 9s.

LATTER-DAY PAMPHLETS. 9s.

LIFE OF JOHN STERLING. With Portrait, 9s.

HISTORY OF FREDERICK THE SECOND. 10 vols., each 9s.

TRANSLATIONS FROM THE GERMAN. 3 vols., each 9s.

EARLY KINGS OF NORWAY; ESSAY ON THE PORTRAITS OF JOHN KNOX; AND GENERAL INDEX. With Portrait Illustrations. 8vo, cloth, 9s.

CHEAP AND UNIFORM EDITION.
23 vols., Crown 8vo, cloth, £7 5s.

THE FRENCH REVOLUTION:
A History. 2 vols., 12s.

OLIVER CROMWELL'S LETTERS AND SPEECHES, with Elucidations, &c. 3 vols., 18s.

LIVES OF SCHILLER AND JOHN STERLING. 1 vol., 6s.

CRITICAL AND MISCELLANEOUS ESSAYS. 4 vols., £1 4s.

SARTOR RESARTUS AND LECTURES ON HEROES. 1 vol., 6s.

LATTER-DAY PAMPHLETS. 1 vol., 6s.

CHARTISM AND PAST AND PRESENT. 1 vol., 6s.

TRANSLATIONS FROM THE GERMAN OF MUSÆUS, TIECK, AND RICHTER. 1 vol., 6s.

WILHELM MEISTER, by Göethe. A Translation. 2 vols., 12s.

HISTORY OF FRIEDRICH THE SECOND, called Frederick the Great. 7 vols., £3 9s.

PEOPLE'S EDITION.
37 vols., small crown 8vo, 37s.; separate vols., 1s. each.

SARTOR RESARTUS. With Portrait of Thomas Carlyle.

FRENCH REVOLUTION. A History. 3 vols.

OLIVER CROMWELL'S LETTERS AND SPEECHES. 5 vols. With Portrait of Oliver Cromwell.

ON HEROES AND HERO WORSHIP, AND THE HEROIC IN HISTORY.

PAST AND PRESENT.

CRITICAL AND MISCELLANEOUS ESSAYS. 4 vols.

THE LIFE OF SCHILLER, AND EXAMINATION OF HIS WORKS. With Portrait.

LATTER-DAY PAMPHLETS.

WILHELM MEISTER. 3 vols.

LIFE OF JOHN STERLING. With Portrait.

HISTORY OF FREDERICK THE GREAT. 10 vols.

TRANSLATIONS FROM MUSÆUS, TIECK, AND RICHTER. 2 vols.

THE EARLY KINGS OF NORWAY, Essay on the Portraits of Knox.

Sets, 37 vols. in 18, 37s.

CHEAP ISSUE.

THE FRENCH REVOLUTION. Complete in 1 vol. With Portrait. Crown 8vo, 2s.

SARTOR RESARTUS, HEROES AND HERO WORSHIP, PAST AND PRESENT, AND CHARTISM. Complete in 1 vol. Crown 8vo, 2s.

OLIVER CROMWELL'S LETTERS AND SPEECHES. Crown 8vo, 2s. 6d.

CRITICAL AND MISCELLANEOUS ESSAYS. 2 vols. 4s.

SIXPENNY EDITION.
4to, sewed.

SARTOR RESARTUS. Eightieth Thousand.

HEROES AND HERO WORSHIP.

ESSAYS: BURNS, JOHNSON, SCOTT, THE DIAMOND NECKLACE.

The above in 1 vol., cloth, 2s. 6d.

DICKENS'S (CHARLES) WORKS.
ORIGINAL EDITIONS.
In demy 8vo.

THE MYSTERY OF EDWIN DROOD. With Illustrations by S. L. Fildes, and a Portrait engraved by Baker. Cloth, 7s. 6d.

OUR MUTUAL FRIEND. With Forty Illustrations by Marcus Stone. Cloth, £1 1s.

THE PICKWICK PAPERS. With Forty-three Illustrations by Seymour and Phiz. Cloth, £1 1s.

NICHOLAS NICKLEBY. With Forty Illustrations by Phiz. Cloth, £1 1s.

SKETCHES BY "BOZ." With Forty Illustrations by George Cruikshank. Cloth, £1 1s.

MARTIN CHUZZLEWIT. With Forty Illustrations by Phiz. Cloth, £1 1s.

DOMBEY AND SON. With Forty Illustrations by Phiz. Cloth, £1 1s.

DAVID COPPERFIELD. With Forty Illustrations by Phiz. Cloth, £1 1s.

BLEAK HOUSE. With Forty Illustrations by Phiz. Cloth, £1 1s.

LITTLE DORRIT. With Forty Illustrations by Phiz. Cloth, £1 1s.

THE OLD CURIOSITY SHOP. With Seventy-five Illustrations by George Cattermole and H. K. Browne. A New Edition. Uniform with the other volumes, £1 1s.

BARNABY RUDGE: a Tale of the Riots of 'Eighty. With Seventy-eight Illustrations by George Cattermole and H. K. Browne. Uniform with the other volumes, £1 1s.

CHRISTMAS BOOKS: Containing—The Christmas Carol; The Cricket on the Hearth; The Chimes; The Battle of Life; The Haunted House. With all the original Illustrations. Cloth, 12s.

OLIVER TWIST and TALE OF TWO CITIES. In one volume. Cloth, £1 1s.

OLIVER TWIST. Separately. With Twenty-four Illustrations by George Cruikshank. Cloth, 11s.

A TALE OF TWO CITIES. Separately. With Sixteen Illustrations by Phiz. Cloth, 9s.

⁎ *The remainder of Dickens's Works were not originally printed in demy 8vo.*

DICKENS'S (CHARLES) WORKS.—*Continued.*
LIBRARY EDITION.
In post 8vo. With the Original Illustrations, 30 vols., cloth, £12.

			s.	d.
PICKWICK PAPERS	43 Illustrns.,	2 vols.	16	0
NICHOLAS NICKLEBY	39 ,,	2 vols.	16	0
MARTIN CHUZZLEWIT	40 ,,	2 vols.	16	0
OLD CURIOSITY SHOP & REPRINTED PIECES	36 ,,	2 vols.	16	0
BARNABY RUDGE and HARD TIMES	36 ,,	2 vols.	16	0
BLEAK HOUSE	40 ,,	2 vols.	16	0
LITTLE DORRIT	40 ,,	2 vols.	16	0
DOMBEY AND SON	38 ,,	2 vols.	16	0
DAVID COPPERFIELD	38 ,,	2 vols.	16	0
OUR MUTUAL FRIEND	40 ,,	2 vols.	16	0
SKETCHES BY "BOZ"	39 ,,	1 vol.	8	0
OLIVER TWIST	24 ,,	1 vol.	8	0
CHRISTMAS BOOKS	17 ,,	1 vol.	8	0
A TALE OF TWO CITIES	16 ,,	1 vol.	8	0
GREAT EXPECTATIONS	8 ,,	1 vol.	8	0
PICTURES FROM ITALY & AMERICAN NOTES	8 ,,	1 vol.	8	0
UNCOMMERCIAL TRAVELLER	8 ,,	1 vol.	8	0
CHILD'S HISTORY OF ENGLAND	8 ,,	1 vol.	8	0
EDWIN DROOD and MISCELLANIES	12 ,,	1 vol.	8	0
CHRISTMAS STORIES from "Household Words," &c.	14 ,,	1 vol.	8	0
THE LIFE OF CHARLES DICKENS. By JOHN FORSTER. With Illustrations. Uniform with this Edition. 10s. 6d.				

A NEW EDITION OF ABOVE, WITH THE ORIGINAL ILLUSTRATIONS, IN LARGE CROWN 8vo, 30 VOLS. IN SETS ONLY.

THE "CHARLES DICKENS" EDITION.
In Crown 8vo. In 21 vols., cloth, with Illustrations, £3 16s.

			s.	d.
PICKWICK PAPERS	8 Illustrations		4	0
MARTIN CHUZZLEWIT	8 ,,		4	0
DOMBEY AND SON	8 ,,		4	0
NICHOLAS NICKLEBY	8 ,,		4	0
DAVID COPPERFIELD	8 ,,		4	0
BLEAK HOUSE	8 ,,		4	0
LITTLE DORRIT	8 ,,		4	0
OUR MUTUAL FRIEND	8 ,,		4	0
BARNABY RUDGE	8 ,,		3	6
OLD CURIOSITY SHOP	8 ,,		3	6
A CHILD'S HISTORY OF ENGLAND	4 ,,		3	6
EDWIN DROOD and OTHER STORIES	8 ,,		3	6
CHRISTMAS STORIES, from "Household Words"	8 ,,		3	6
SKETCHES BY "BOZ"	8 ,,		3	6
AMERICAN NOTES and REPRINTED PIECES	8 ,,		3	6
CHRISTMAS BOOKS	8 ,,		3	6
OLIVER TWIST	8 ,,		3	6
GREAT EXPECTATIONS	8 ,,		3	6
TALE OF TWO CITIES	8 ,,		3	0
HARD TIMES and PICTURES FROM ITALY	8 ,,		3	0
UNCOMMERCIAL TRAVELLER	4 ,,		3	0
THE LIFE OF CHARLES DICKENS. Numerous Illustrations.		2 vols.	7	0
THE LETTERS OF CHARLES DICKENS		2 vols.	7	0

DICKENS'S (CHARLES) WORKS.—*Continued.*

THE ILLUSTRATED LIBRARY EDITION.
(WITH LIFE.)

Complete in 32 Volumes. Demy 8vo, 10s. each; or set, £16.

This Edition is printed on a finer paper and in a larger type than has been employed in any previous edition. The type has been cast especially for it, and the page is of a size to admit of the introduction of all the original illustrations.

No such attractive issue has been made of the writings of Mr. Dickens, which, various as have been the forms of publication adapted to the demands of an ever widely-increasing popularity, have never yet been worthily presented in a really handsome library form.

The collection comprises all the minor writings it was Mr. Dickens's wish to preserve.

SKETCHES BY "BOZ." With 40 Illustrations by George Cruikshank.
PICKWICK PAPERS. 2 vols. With 42 Illustrations by Phiz.
OLIVER TWIST. With 24 Illustrations by Cruikshank.
NICHOLAS NICKLEBY. 2 vols. With 40 Illustrations by Phiz.
OLD CURIOSITY SHOP and REPRINTED PIECES. 2 vols. With Illustrations by Cattermole, &c.
BARNABY RUDGE and HARD TIMES. 2 vols. With Illustrations by Cattermole, &c.
MARTIN CHUZZLEWIT. 2 vols. With 40 Illustrations by Phiz.
AMERICAN NOTES and PICTURES FROM ITALY. 1 vol. With 8 Illustrations.
DOMBEY AND SON. 2 vols. With 40 Illustrations by Phiz.
DAVID COPPERFIELD. 2 vols. With 40 Illustrations by Phiz.
BLEAK HOUSE. 2 vols. With 40 Illustrations by Phiz.
LITTLE DORRIT. 2 vols. With 40 Illustrations by Phiz.
A TALE OF TWO CITIES. With 16 Illustrations by Phiz.
THE UNCOMMERCIAL TRAVELLER. With 8 Illustrations by Marcus Stone.
GREAT EXPECTATIONS. With 8 Illustrations by Marcus Stone.
OUR MUTUAL FRIEND. 2 vols. With 40 Illustrations by Marcus Stone.
CHRISTMAS BOOKS. With 17 Illustrations by Sir Edwin Landseer, R.A., Maclise, R.A., &c. &c.
HISTORY OF ENGLAND. With 8 Illustrations by Marcus Stone.
CHRISTMAS STORIES. (From "Household Words" and "All the Year Round.") With 14 Illustrations.
EDWIN DROOD AND OTHER STORIES. With 12 Illustrations by S. L. Fildes.
LIFE OF CHARLES DICKENS. By John Forster. With Portraits. 2 vols. (not separate.)

DICKENS'S (CHARLES) WORKS.—*Continued.*

THE POPULAR LIBRARY EDITION
OF THE WORKS OF
CHARLES DICKENS,

In 30 *Vols., large crown* 8*vo, price* £6; *separate Vols.* 4*s. each.*

An Edition printed on good paper, each volume containing 16 full-page Illustrations, selected from the Household Edition, on Plate Paper.

SKETCHES BY "BOZ."
PICKWICK. 2 vols.
OLIVER TWIST.
NICHOLAS NICKLEBY 2 vols.
MARTIN CHUZZLEWIT. 2 vols.
DOMBEY AND SON. 2 vols.
DAVID COPPERFIELD. 2 vols.
CHRISTMAS BOOKS.
OUR MUTUAL FRIEND. 2 vols.
CHRISTMAS STORIES.
BLEAK HOUSE. 2 vols.
LITTLE DORRIT. 2 vols.
OLD CURIOSITY SHOP AND REPRINTED PIECES. 2 vols
BARNABY RUDGE. 2 vols.
UNCOMMERCIAL TRAVELLER.
GREAT EXPECTATIONS.
TALE OF TWO CITIES.
CHILD'S HISTORY OF ENGLAND.
EDWIN DROOD AND MISCELLANIES.
PICTURES FROM ITALY AND AMERICAN NOTES.

DICKENS'S (CHARLES) WORKS.—*Continued.*

HOUSEHOLD EDITION.

(WITH LIFE.)

In 22 Volumes. Crown 4to, cloth, £4 8s. 6d.

MARTIN CHUZZLEWIT, with 59 Illustrations, cloth, 5s.

DAVID COPPERFIELD, with 60 Illustrations and a Portrait, cloth, 5s.

BLEAK HOUSE, with 61 Illustrations, cloth, 5s.

LITTLE DORRIT, with 58 Illustrations, cloth, 5s.

PICKWICK PAPERS, with 56 Illustrations, cloth, 5s.

OUR MUTUAL FRIEND, with 58 Illustrations, cloth, 5s.

NICHOLAS NICKLEBY, with 59 Illustrations, cloth, 5s.

DOMBEY AND SON, with 61 Illustrations, cloth, 5s.

EDWIN DROOD; REPRINTED PIECES; and other Stories, with 30 Illustrations, cloth, 5s.

THE LIFE OF DICKENS. By JOHN FORSTER. With 40 Illustrations. Cloth, 5s.

BARNABY RUDGE, with 46 Illustrations, cloth, 4s.

OLD CURIOSITY SHOP, with 32 Illustrations, cloth, 4s.

CHRISTMAS STORIES, with 23 Illustrations, cloth, 4s.

OLIVER TWIST, with 28 Illustrations, cloth, 3s.

GREAT EXPECTATIONS, with 26 Illustrations, cloth, 3s.

SKETCHES BY "BOZ," with 36 Illustrations, cloth, 3s.

UNCOMMERCIAL TRAVELLER, with 26 Illustrations, cloth, 3s.

CHRISTMAS BOOKS, with 28 Illustrations, cloth, 3s.

THE HISTORY OF ENGLAND, with 15 Illustrations, cloth, 3s.

AMERICAN NOTES and PICTURES FROM ITALY, with 18 Illustrations cloth, 3s.

A TALE OF TWO CITIES, with 25 Illustrations, cloth, 3s.

HARD TIMES, with 20 Illustrations, cloth, 2s. 6d.

DICKENS'S (CHARLES) WORKS.—*Continued.*

THE CABINET EDITION.

In 32 vols. small fcap. 8vo, Marble Paper Sides, Cloth Backs, with uncut edges, price Eighteenpence each.

Each Volume contains Eight Illustrations reproduced from the Originals.

CHRISTMAS BOOKS.
MARTIN CHUZZLEWIT, Two Vols.
DAVID COPPERFIELD, Two Vols.
OLIVER TWIST.
GREAT EXPECTATIONS.
NICHOLAS NICKLEBY, Two Vols.
SKETCHES BY "BOZ."
CHRISTMAS STORIES.
THE PICKWICK PAPERS, Two Vols.
BARNABY RUDGE, Two Vols.
BLEAK HOUSE, Two Vols.
AMERICAN NOTES AND PICTURES FROM ITALY.
EDWIN DROOD; AND OTHER STORIES.
THE OLD CURIOSITY SHOP, Two Vols.
A CHILD'S HISTORY OF ENGLAND
DOMBEY AND SON, Two Vols.
A TALE OF TWO CITIES.
LITTLE DORRIT, Two Vols.
MUTUAL FRIEND, Two Vols.
HARD TIMES.
UNCOMMERCIAL TRAVELLER.
REPRINTED PIECES.

NEW & CHEAP ISSUE OF THE WORKS OF CHARLES DICKENS.

In Pocket Volumes.

PICKWICK PAPERS, with 8 Illustrations, cloth, 2s.
NICHOLAS NICKLEBY, with 8 Illustrations, cloth, 2s.
OLIVER TWIST, with 8 Illustrations, cloth, 1s.
SKETCHES BY "BOZ," with 8 Illustrations, cloth, 1s.
OLD CURIOSITY SHOP, with 8 Illustrations, cloth, 2s.
BARNABY RUDGE, with 16 Illustrations, cloth, 2s.
AMERICAN NOTES AND PICTURES FROM ITALY, with 8 Illustrations, cloth, 1s. 6d.
CHRISTMAS BOOKS, with 8 Illustrations, cloth, 1s. 6d.
MARTIN CHUZZLEWIT, with 8 Illustrations, 2s.

DICKENS'S (CHARLES) WORKS.—*Continued.*

MR. DICKENS'S READINGS.
Fcap. 8vo, sewed.

CHRISTMAS CAROL IN PROSE. 1s.
CRICKET ON THE HEARTH. 1s.
CHIMES: A GOBLIN STORY. 1s.
STORY OF LITTLE DOMBEY. 1s.
POOR TRAVELLER, BOOTS AT THE HOLLY-TREE INN, and MRS. GAMP. 1s.

A CHRISTMAS CAROL, with the Original Coloured Plates.
Being a reprint of the Original Edition. With red border lines. Small 8vo, red cloth, gilt edges, 5s.

CHARLES DICKENS'S CHRISTMAS BOOKS.
REPRINTED FROM THE ORIGINAL PLATES.
Illustrated by JOHN LEECH, D. MACLISE, R.A., R. DOYLE, C. STANFIELD, R.A., &c.
Fcap. cloth, 1s. each. Complete in a case, 5s.

A CHRISTMAS CAROL IN PROSE.
THE CHIMES: A Goblin Story.
THE CRICKET ON THE HEARTH: A Fairy Tale of Home.
THE BATTLE OF LIFE. A Love Story.
THE HAUNTED MAN AND THE GHOST'S STORY.

SIXPENNY REPRINTS.

READINGS FROM THE WORKS OF CHARLES DICKENS.
As selected and read by himself and now published for the first time. Illustrated

A CHRISTMAS CAROL, AND THE HAUNTED MAN. By CHARLES DICKENS. Illustrated.

THE CHIMES: A GOBLIN STORY, AND THE CRICKET ON THE HEARTH. Illustrated.

THE BATTLE OF LIFE: A LOVE STORY, HUNTED DOWN, AND A HOLIDAY ROMANCE. Illustrated.

The last Three Volumes as Christmas Works, In One Volume, red cloth, 2s. 6d.

SCIENCE AND ART,

A Journal for Teachers and Students.

ISSUED BY Messrs. CHAPMAN & HALL, Limited,

Agents for the Science and Art Department of the Committee of Council on Education.

MONTHLY, PRICE THREEPENCE.

The Journal contains contributions by distinguished men; short papers by prominent teachers; leading articles; correspondence; answers to questions set at the May Examinations of the Science and Art Department; and interesting news in connection with the scientific and artistic world.

PRIZE COMPETITION.

With each issue of the Journal, papers or drawings are offered for Prize Competition, extending over the range of subjects of the Science and Art Department and City and Guilds of London Institute.

There are thousands of Science and Art Schools and Classes in the United Kingdom, but the teachers connected with these institutions, although engaged in the advancement of identical objects, are seldom known to each other except through personal friendship. One object of the new Journal is to enable those engaged in this common work to communicate upon subjects of importance, with a view to an interchange of ideas, and the establishment of unity of action in the various centres.

TERMS OF SUBSCRIPTION.

ONE YEAR'S SUBSCRIPTION	3s. 0d.
HALF ,, ,, ...	1s. 6d.
SINGLE COPY	3d.
POSTAGE MONTHLY EXTRA	1d.

Cheques and Post Office Orders to be made payable to
CHAPMAN & HALL, Limited.

ANSWERS TO QUESTIONS, 1887 and 1888.

Messrs. CHAPMAN & HALL beg to announce that Answers to the Questions (Elementary and Advanced) set at the Examinations of the Science and Art Department of May, 1887 and 1888, are published as under, each subject being kept distinct, and issued in pamphlet form separately.

1. ANIMAL PHYSIOLOGY ... 1887 — By J. H. E. Brock, M.D., B.S. (Lond.),
   ,, ,, 1888 — F.R.C.S. (Eng.), D.P.H. (Lond.)
2. BUILDING CONSTRUCTION 1887 } H. Adams, C.E., M.I.M.E.
   ,, ,, 1888 }
3. THEORETICAL MECHANICS, 1887 — J. C. Fell, M.I.M.E.
   ,, ,, 1888 — E. Pillow, M.I.M.E.
4. INORGANIC CHEMISTRY (Theoretical), 1887 — Rev. F. W. Harnett, M.A.
   INORGANIC CHEMISTRY (Theoretical), 1888 — J. J. Pilley, Ph.D., F.C.S., F.R.M.S.
5. Ditto—ALTERNATIVE COURSE 1887 } J. Howard, F.C.S.
   Ditto—ALTERNATIVE COURSE 1888 }
6. MAGNETISM AND ELECTRICITY 1887 } W. Hibbert, F.I.C., A.I.E.E.
   MAGNETISM AND ELECTRICITY 1888 }
7. PHYSIOGRAPHY ... ... 1887 } W. Rheam, B.Sc.
   ,, ,, ... 1888 }
8. PRACTICAL PLANE AND SOLID GEOMETRY ... ... 1887 } H. Angel.
   PRACTICAL PLANE AND SOLID GEOMETRY ... ... 1888 }
9. ART—THIRD GRADE. PERSPECTIVE .. ... 1887 — A. Fisher.
   ART—THIRD GRADE. PERSPECTIVE ... 1888 — A. Fisher and S. Beale.
10. PURE MATHEMATICS ... 1887 — R. R. Steel, F.C.S.
    ,, ,, ... 1888 — H. Carter, B.A.
11. MACHINE CONSTRUCTION AND DRAWING ... ... 1887 } H. Adams, C.E., M.I.M.E.
    MACHINE CONSTRUCTION AND DRAWING ... ... 1888 }
12. PRINCIPLES OF AGRICULTURE 1887 } Dr. H. J. Webb, B.Sc.
    PRINCIPLES OF AGRICULTURE 1888 }
13. SOUND, LIGHT, AND HEAT, 1887 } C. A. Stevens.
    ,, ,, ,, 1888 }
14. HYGIENE ... ... 1887 } J. J. Pilley, Ph.D., F.C.S., F.R.M.S.
    ,, ... 1888 }
15. INORGANIC CHEMISTRY (Practical) ... ... 1887 } J. Howard, F.C.S.
    INORGANIC CHEMISTRY (Practical) ... ... 1888 }
16. APPLIED MECHANICS ... 1888 — C. B. Outon, Wh.Sc.

The price of each Pamphlet (dealing with both Elementary and Advanced papers) will be 2d. net, postage included. Special terms will be given if quantities are ordered.

THE FORTNIGHTLY REVIEW.
Edited by FRANK HARRIS.

THE FORTNIGHTLY REVIEW is published on the 1st of every month, and a Volume is completed every Six Months.

The following are among the Contributors:—

ADMIRAL LORD ALCESTER.
GRANT ALLEN.
SIR RUTHERFORD ALCOCK.
AUTHOR OF "GREATER BRITAIN."
PROFESSOR BAIN.
SIR SAMUEL BAKER.
PROFESSOR BEESLY.
PAUL BOURGET.
BARON GEORGE VON BUNSEN.
DR. BRIDGES.
HON. GEORGE C. BRODRICK.
JAMES BRYCE, M.P.
THOMAS BURT, M.P.
SIR GEORGE CAMPBELL, M.P.
THE EARL OF CARNARVON.
EMILIO CASTELAR.
RT. HON. J. CHAMBERLAIN, M.P.
PROFESSOR SIDNEY COLVIN.
THE EARL COMPTON.
MONTAGUE COOKSON, Q.C.
L. H. COURTNEY, M.P.
G. H. DARWIN.
SIR GEORGE W. DASENT.
PROFESSOR A. V. DICEY.
PROFESSOR DOWDEN.
RT. HON. M. E. GRANT DUFF.
RIGHT HON. H. FAWCETT, M.P.
ARCHDEACON FARRAR.
EDWARD A. FREEMAN.
J. A. FROUDE.
MRS. GARRET-ANDERSON.
J. W. L. GLAISHER, F.R.S.
SIR J. E. GORST, Q.C., M.P.
EDMUND GOSSE.
THOMAS HARE.
FREDERIC HARRISON.
ADMIRAL SIR G. P. HORNBY.
LORD HOUGHTON.
PROFESSOR HUXLEY.
PROFESSOR R. C. JEBB.
ANDREW LANG.
ÉMILE DE LAVELEYE.
T. E. CLIFFE LESLIE.
W. S. LILLY.
MARQUIS OF LORNE.

PIERRE LOTE.
SIR JOHN LUBBOCK, BART., M.P.
THE EARL OF LYTTON.
SIR H. S. MAINE.
CARDINAL MANNING.
DR. MAUDSLEY.
PROFESSOR MAX MÜLLER.
GEORGE MEREDITH.
RT. HON. G. OSBORNE MORGAN, Q.C., M.P.
PROFESSOR HENRY MORLEY.
RT. HON. JOHN MORLEY, M.P.
WILLIAM MORRIS.
PROFESSOR H. N. MOSELEY.
F. W. H. MYERS.
F. W. NEWMAN.
PROFESSOR JOHN NICHOL.
W. G. PALGRAVE.
WALTER H. PATER.
RT. HON. LYON PLAYFAIR, M.P.
SIR HENRY POTTINGER, BART.
PROFESSOR J. R. SEELEY.
LORD SHERBROOKE.
PROFESSOR SIDGWICK.
HERBERT SPENCER.
M. JULES SIMON.
  (DOCTOR L'ACADEMIE FRANCAISE)
HON. E. L. STANLEY.
SIR J. FITZJAMES STEPHEN, Q.C.
LESLIE STEPHEN.
J. HUTCHISON STIRLING.
A. C. SWINBURNE.
DR. VON SYBEL.
J. A. SYMONDS.
SIR THOMAS SYMONDS.
  (ADMIRAL OF THE FLEET).
THE REV. EDWARD F. TALBOT
  (WARDEN OF KEBLE COLLEGE).
SIR RICHARD TEMPLE, BART.
HON. LIONEL A. TOLLEMACHE.
H. D. TRAILL.
PROFESSOR TYNDALL.
A. J. WILSON.
GEN. VISCOUNT WOLSELEY.
THE EDITOR.

&c. &c. &c.

THE FORTNIGHTLY REVIEW *is published at* 2s. 6d.

CHAPMAN & HALL, LIMITED, 11, HENRIETTA STREET, COVENT GARDEN, W.C.

CHARLES DICKENS AND EVANS,] [CRYSTAL PALACE PRESS.

www.ingramcontent.com/pod-product-compliance
Lightning Source LLC
Chambersburg PA
CBHW022106290426
44112CB00008B/563